Communications
in Computer and Information Science 926

Commenced Publication in 2007
Founding and Former Series Editors:
Phoebe Chen, Alfredo Cuzzocrea, Xiaoyong Du, Orhun Kara, Ting Liu,
Dominik Ślęzak, and Xiaokang Yang

More information about this series at http://www.springer.com/series/7899

Jonice Oliveira · Claudio M. Farias
Esther Pacitti · Giancarlo Fortino (Eds.)

Big Social Data
and Urban Computing

First Workshop, BiDU 2018
Rio de Janeiro, Brazil, August 31, 2018
Revised Selected Papers

 Springer

Editors
Jonice Oliveira (iD)
Universidade Federal do Rio de Janeiro
Rio de Janeiro, Rio de Janeiro, Brazil

Esther Pacitti
Inria/CNRS
University of Montpellier
Montpellier, France

Claudio M. Farias
Universidade Federal do Rio de Janeiro
Rio de Janeiro, Brazil

Giancarlo Fortino
University of Calabria (Unical)
Rende (CS), Italy

ISSN 1865-0929   ISSN 1865-0937   (electronic)
Communications in Computer and Information Science
ISBN 978-3-030-11237-0   ISBN 978-3-030-11238-7   (eBook)
https://doi.org/10.1007/978-3-030-11238-7

Library of Congress Control Number: 2018967045

This Springer imprint is published by the registered company Springer Nature Switzerland AG
The registered company address is: Gewerbestrasse 11, 6330 Cham, Switzerland

BiDU – Workshop on Big Social Data and Urban Computing

In urban spaces, there is a huge amount of heterogeneous data being generated by a diversity of sources, such as sensors, devices, vehicles, smart buildings, and others. Although they are used to monitor basic services, they can provide significant information about human interactions and populational dynamics. Moreover, people constantly interact with each other through social media services, and much of interpersonal interaction is nowadays mediated by information technology. Citizens consume and share information about their cities—such as problems, events, ideas, suggestions, criticisms, and demands—acting as "human sensors," forming opinions and participating in the city evolution.

This data explosion has resulted in the emerging topic of "Big Social Data." Broadly speaking, big social data refers to large data volumes that relate to people interactions or describe their behaviors, needs, and patterns. The volume, the production and spreading velocity, and the variety (providing semantic richness) of such data open up enormous possibilities to utilize and analyze them for the understanding of urban spaces, tackling the major issues that these localities face, and helping in the creation of smarter and sustainable cities.

Urban computing is a process of acquisition, treatment, and analysis of big and heterogeneous data to better understand how city ecosystems work. This understanding can remedy a wide range of issues affecting the everyday lives of citizens and the long-term health and efficiency of cities. The use of big social data in urban computing helps us to understand the nature of urban phenomena and even predict the future of cities, creating solutions to reduce costs and optimize resource consumption, improve population mobility, provide higher human life quality, enhance decision-making in emergency scenarios, and engage more effectively with citizens for a continuous city planning.

This workshop, which was held in conjunction with 44th International Conference on Very Large Data Bases (VLDB), in Rio de Janeiro, and connected works about the use and treatment of big social data in multidisciplinary research spanning across computer science. All papers went through a double-blind review process, with at least three reviewers, and were reviewed according to the following criteria: adequacy of workshop scope, relevance, technical quality, clarity, originality, and evaluation of results. The papers were categorized in: full papers (Research, Experiments and Case Studies, Industry and Application, and Dataset papers) and short papers (Vision Papers). We received 40 submissions (full papers: 30, vision papers: 10), of which we selected 11 full papers and 16 posters. All the full papers were orally presented and distributed in the sections: Session 1 – Urban Mobility, Session 2 – Urban Sensing, Session 3 – Contemporary Social Problems, and Session 4 – Collaboration and Crowdsourcing.

Moreover, the BiDU workshop had a keynote entitled "Landscape of Practical Blockchain Systems and Their Applications" by Dr. C. Mohan (IBM Almaden Research Center & Tsinghua University). Also, the panel "Social Computing for Smarter Cities" featured Sihem Amer-Yahia (Laboratoire d'Informatique de Grenoble), Elaine Rabello (FIOCRUZ and Universidade do Estado do Rio de Janeiro), Gabriela Ruberg (Central Bank of Brazil), Bill Howe (University of Washington), and moderated by Mirella Moro (Universidade Federal de Minas Gerais). We thank these inspiring speakers for accepting our invitation and enlightening this event.

We would like to sincerely thank the VLDB organization for the constant help and support, the Program Committee members and reviewers for their excellent work and invaluable support during the review process, and the authors of the submitted papers for their very interesting and high-quality contributions. Most importantly, we thank all the attendees who ensured BiDU could be an appropriate forum where the sharing of knowledge and experiences of big social data and urban computing promoted new advances in both research and development. We also thank the Springer team, especially Mr. Jorge Nakahara.

We hope that readers enjoy the papers included in the proceedings.

December 2018

Jonice Oliveira
Claudio M. Farias
Esther Pacitti
Giancarlo Fortino

Organization

Program Chairs

Jonice Oliveira   Universidade Federal do Rio de Janeiro, Brazil
Claudio Miceli   Universidade Federal do Rio de Janeiro, Brazil
Esther Pacitti   Inria/CNRS, University of Montpellier, France
Giancarlo Fortino   Università della Calabria, Italy

Program Committee

Adriano Pereira	Universidade Federal de Minas Gerais, Brazil
Ahmed Elmisery	Universidad Técnica Federico Santa María, Chile
Antonio Liota	Technische Universiteit Eindhoven, Netherlands
Arthur Zivianni	Laboratório Nacional de Computação Científica, Brazil
Carlos Sarraute	Instituto Tecnológico de Buenos Aires, Argentina
Chiara Renso	ISTI Institute of CNR, Italy
Chico Camargo	University of Oxford, UK
Claudio Miceli	Universidade Federal do Rio de Janeiro, Brazil
Eduardo Ogasawara	Centro Federal de Educação Tecnológica Celso Suckow da Fonseca, Brazil
Elisabeth Lex	Graz University of Technology, Know-Center, Austria
Esther Pacitti	Inria/CNRS, University of Montpellier, France
Flavia Bernardini	Universidade Federal Fluminense, Brazil
Flávia Coimbra Delicato	Universidade Federal do Rio de Janeiro, Brazil
Giancarlo Fortino	Università della Calabria, Italy
Giseli Rabello Lopes	Universidade Federal do Rio de Janeiro, Brazil
Giuseppe Di Fatta	University of Reading, UK
Grazziela Figueredo	University of Nottingham, UK
Haibin Zhu	Nipissing University, Canada
Igor Santos	Centro Federal de Educação Tecnológica Celso Suckow da Fonseca, Brazil
Javier Baliosian	Universidada de la República, Uruguay
Jonice Oliveira	Universidade Federal do Rio de Janeiro, Brazil
José Viterbo	Universidade Federal Fluminense, Brazil
Juan Antonio Lossio Ventura	University of Florida, USA
Karima Boudaoud	Ecole Polytechnique de l'Université de Nice Sophia Antipolis, France
Lívia Ruback	Universidade Federal do Rio de Janeiro, Brazil
Manel Zarrouk	Insight Centre, NUIG, Ireland
Marcelo Mendoza	Universidad Técnica Federico Santa María, Chile

Marcos Oliveira	GESIS: Leibniz Institute for the Social Sciences, Germany
Maria Luiza Campos	Universidade Federal do Rio de Janeiro, Brazil
Mirella M. Moro	Universidade Federal de Minas Gerais, Brazil
Nazim Agoulmine	Université d'Évry Val d'Essonne, France
Paulo de Figueiredo Pires	Universidade Federal do Rio de Janeiro, Brazil
Reinaldo Bezerra Braga	Instituto Federal do Ceará, Brazil
Reyes Juarez Ramirez	Universidad Autonoma da Baja California, Mexico
Rodrigo de Souza Couto	Universidade Estadual do Rio de Janeiro, Brazil
Rodrigo Santos	Universidade Federal do Estado do Rio de Janeiro, Brazil
Sérgio Lifschitz	PUC-Rio, Brazil
Sergio Ochoa	Universidad de Chile, Chile
Soon Ae Chun	City University of New York, USA
Taniro Rodrigues	Universidade Federal do Rio Grande do Norte, Brazil
Thiago H. Silva	Tecnológica Federal do Paraná, Brazil
Thiago Moreira	Université d'Évry Val d'Essonne, France
Wei Li	University of Sydney, Australia

Additional Reviewers

Andrea Vinci	Istituto di Calcolo e Reti ad Alte Prestazioni/CNR, Italy
Antonio Guerrieri	Università della Calabria, Italy
Carlos Eduardo Barbosa	Universidade Federal do Rio de Janeiro, Brazil
Claudio Savaglio	Università della Calabria, Italy
Danilo Carvalho	Universidade Federal do Rio de Janeiro, Brazil

Contents

Collaboration and Crowdsourcing

Urban Mobility

Characterizing Usage Patterns and Service Demand of a Two-Way Car-Sharing System

Felipe Rooke[1]([✉]) [ID], Victor Aquiles[1] [ID], Alex Borges Vieira[1] [ID],
Jussara M. Almeida[2] [ID], and Idilio Drago[3] [ID]

[1] Departamento de Ciência da Computação, Universidade Federal de Juiz de Fora,
Juiz de Fora, MG, Brazil
{felipe.rooke,alex.borges}@ufjf.edu.br, victoraquiles@ice.ufjf.br
[2] Departamento de Ciência da Computação, Universidade Federal de Minas Gerais,
Belo Horizonte, MG, Brazil
jussara@dcc.ufmg.br
[3] Department of Electronics and Telecommunications, Politecnico di Torino,
Turin, Italy
idilio.drago@polito.it

Abstract. Urban mobility is directly linked to the demand for communication resources and, clearly, its understanding is useful for better planning of urban and communication systems. However, getting data about urban mobility is still a challenge. In many cases, only a few companies have access to accurate and updated data. In most cases, these data are also privacy sensitive. It is thus important to generate models that can help to understand mobility patterns. We here characterize the demands of a two-way car-sharing system. We explore data of the public API of Modo, a car-sharing system that operates in Vancouver (Canada) and nearby regions. Our study uncovers patterns of users' habits and demands in the service, which can be explored for urban and communication planning.

Keywords: Car-sharing · Two-way · Characterizing · Urban mobility · Patterns

1 Introduction

The comprehension of urban mobility has been a target of studies and investments. Urban mobility is a key research area, attracting several academic studies and private investments. It is intrinsically connected to a wide number of urban activities, such as the demand for communication resources. Indeed, the massification of mobile devices turned the network access ubiquitous and user-centered. The actual network infrastructure is each day less rigid, and users demand communication while moving across the city. Understanding the urban mobility, specifically the traffic-related mobility, is important for a series of tasks, ranging

© Springer Nature Switzerland AG 2019
J. Oliveira et al. (Eds.): BiDU 2018, CCIS 926, pp. 3–17, 2019.
https://doi.org/10.1007/978-3-030-11238-7_1

from road mesh planning to communication resources allocation (Herrera et al. 2010; Ma et al. 2013).

The first step in understanding urban mobility patterns is the proper acquisition of data. Data related to this problem domain can be obtained by several ways, e.g., by observing vehicles passing through sensors or fixed/mobile radars, by acquiring traffic data from cameras, or even by the active participation of users (*crowdsourcing*). However, the data acquisition is still a challenge. Only a few companies have access to accurate data, and most of the time these data are privacy sensitive (Ciociola et al. 2017). Therefore, it is important to generate models that can help to understand the urban mobility and the social interactions of people in the urban environment.

Many alternative transport modes contribute to urban mobility. Among them, car-sharing systems have received an increased attention from the academic community (Boldrini et al. 2016; Ciociola et al. 2017; Becker et al. 2017). In a car-sharing system, people can schedule the use of a vehicle, without worrying about maintenance and parking fees. These systems have already a large volume of users and thus can be representative of an important type of urban mobility pattern. In fact, by 2015, more than 1.5 million users and more than 22,000 shared vehicles have been counted in the Americas, and growth in usage is still expected (Shaheen 2016).

There are different business models for operating car-sharing services (Nourinejad 2014): (i) *one-way* services, which rely on base stations scattered in one region, with users renting and returning vehicles at arbitrary stations; (ii) *two-way* services, in which users must return the vehicles always at the same station where the rent started; and (iii) *free-floating* services, where there is no base stations, and users are free to rent and return vehicles at any position inside the operating area of the service (Boldrini et al. 2016).

Recent studies have addressed one-way services, showing spatial-temporal usage characteristics (Boldrini et al. 2016). Similarly, user characteristics and usage patterns in free-floating services have been addressed (Kopp et al. 2015; Ciociola et al. 2017; Cocca et al. 2018). However, there is no study that characterizes and models two-way services.

In this work, we characterize usage patterns and the demands of a two-way car-sharing system. More precisely, we explore the data offered by the public API of *Mode*,[1] a car-sharing service that operates in Vancouver (Canada) and nearby regions. Our contributions are the following:

(i) the characterization of demands in a large two-way car-sharing service;
(ii) the study of the system workloads, which can be exploited, e.g., for planning urban and communication systems.

We believe our study is an important step towards understanding all types of car-sharing usage. It can help uncover particular situations where such services are attractive and, together with data from other transport modes, help to

---

[1] http://www.modo.coop/.

uncover trends and mobility patterns. In fact, the data and its characterization can support decision-making related to urban mobility planning.

The paper is organized as follows: Sect. 2 describes the operation of existing car-sharing services and introduces details of the two-way model, the focus of this work; Sect. 3 discusses the data collection methodology and describes the acquired dataset; Sect. 4 presents results obtained from the characterization and analysis of the dataset; Sect. 5 describes related work, whereas Sect. 6 concludes the paper.

2 Car-Sharing System Basic Model

The first concepts of car-sharing systems date back to 1948. Although, the basic principles of such service were consolidated during the 1970's (Harms and Truffer 1998). At a glance, the key idea behind car-sharing systems is that a fleet of cars can be shared by several users. They drive a car whenever they need without owning the car.

During the 1990's, along with emerging problems of large urban centers, high fuel prices, traffic congestion, high emission of pollutants, the idea of sharing came back again (Becker et al. 2017). Since then, car sharing has been the subject of academy studies (Millard-Ball 2005). Understanding the dynamics of these services provides valuable insights into how people move in urban centers. This information can give support to precise and efficient urban planning, ranging from traffic planning or the design of communication infrastructures.

There are two major car-sharing models: station-based and free-floating. Moreover, station-based may be divided into *one-way services* and two-way services. Station-based models require that a user pick up the vehicle she/he will use at a given base station. The user, in turn, may leave the vehicle at any of the base stations scattered throughout the service coverage region (i.e., one-way car-sharing service), or she/he may be obliged to return the vehicle to the station of origin (i.e., two-way car-sharing service). Clearly, the two-way model requires simpler logistics and infrastructure compared to other models. Its implementation can be performed at a lower cost and higher speed.

Note that, car-sharing, in special the two-way model, differs from classical car rental in many ways. Indeed, car-sharing is a self-service based service, where vehicles can be allocated in fractions of times, as well as by the day, as traditional car rental. Moreover, the one-way model may be more flexible and cost-efficient to users than classical rental. For example, in case there is a base station near to the final user destination, she/he may leave the car at the station while she/he perform other tasks. The time the vehicle is parked is not charged, incurring to lower costs to users. However, users may not be able to make a new reservation, in case the same vehicle is reserved by another user.

The free-floating model does not require any fixed station. In other words, users reserve the nearest car, parked into city streets. By the end of the use, users may leave vehicles at any location in a predefined area. Notably, free-floating model eliminates the limitations that station-based models hold, making the experience more flexible and closer personal vehicles (Ciari et al. 2014). However, there are some problems, such as the uncertainty of finding nearby cars.

We are aware of works that characterize and model one-way services (Ciari et al. 2014; Stillwater et al. 2009; Burkhardt and Millard-Ball 2006). These works achieve a consensus about some questions as: (i) the most accepted markets for these services are in dense urban areas with good public transport (Stillwater et al. 2009); (ii) the profile of the users of these systems is composed of young people with high income and good schooling (Burkhardt and Millard-Ball 2006). A number of researchers also confirm positive impacts on the actual transport system, such as the reducing on traffic and emission of pollutants (Cervero and Tsai 2004; Martin and Shaheen 2011), the reducing on parking areas and the increase in the use of public transport (Shaheen et al. 2010). However, to the best of our knowledge, there are no characterization studies or models for two-way car-sharing services and so there is a gap to be explored.

Figure 1 presents an abstract model that describes the operation of a car-sharing system. This simple model can be applied to three types of services. Note that there are four possible states for a vehicle: available, partially available, rented, and unavailable. These states define when a vehicle is busy or idle. A vehicle, when reserved, passes from the available state to the partially available state (1). The reservation can also occur as an immediate rent, in that case passing from the state available to rent (2). From the partially available state, when the reserved time arrives and the rental starts, the vehicle moves to the rented state (3). In the case of cancellation of a reservation, the vehicle returns to the available state (4). Starting from the rented state, a vehicle may return to the partially available state, when a rental is finalized and there is a reservation scheduled in the interval (5) or it may return to the available state when there is no more reserve for it (6). The vehicle enters the unavailable state when it is in maintenance or out of service (7), and it returns to the system when issues have been resolved (8).

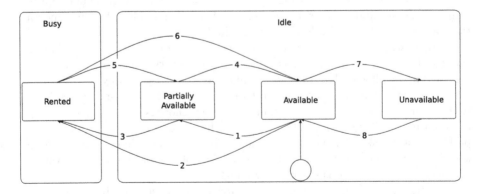

Fig. 1. States of a vehicle in a car-sharing system.

3   Dataset and Methodology

Our study relies on data crawled from the public API of Modo, a car-sharing service operating in Vancouver (Canada) and nearby regions. Modo, in 2017, featured about 600 vehicles, distributed among conventional, electric and hybrid cars. In addition, this system covers about 18000 users, over an area of approximately $133\,km^2$.

The data collection process was conducted using a crawler which uses the API provided by the Modo[2] platform. The crawling process allows us to gather data about vehicles available on the platform. These data enable to study the distribution of supply and demand for vehicles in time and space.

The first step was to request the API a list of all vehicles of the platform. We perform a request per minute, for every available vehicle, and each request to the Modo API returns the period a specific vehicle will be available, during the next 24-h interval. In addition, the responses include the vehicle location and its station identification. We discarded data from unavailable vehicles, i.e., in maintenance or out of service.

Note that Modo API does not return specific vehicle status, nor any information that could be used to identify users of the system. We uncover if a vehicle is busy or idle based on its reservation period and the current observation time. For example, Fig. 2 illustrates the process of collecting data for a given vehicle.

As we previous stated, each API request tells us the period a given vehicle is booked in the next 24-h. According to Fig. 2, we leverage three possible situations:

- First, as shown in Fig. 2a, at $t = 1$, we perform a request to the Modo API and note that a given vehicle is booked between $t = [1; 5]$. At $t = 2$, the new request to the Modo API still returning the previously booking period. Each following request to the API confirms the booking period. In this case, at time $t = 6$, we perform a request to the API and the vehicle is no longer booked. In sum, we are able to judge that someone booked the vehicle between $t = [1; 5]$, used the car and, disposed of it after the original booking period.
- Second, as shown in Fig. 2b, at $t = 1$, we perform a request to the Modo API and note that a given vehicle is booked between $t = [1; 6]$. The request at $t = 2$ still confirms the previously booking period. However, in this case, a request at $t = 3$ shows no booking period at this moment. In this case, we judge that user canceled the future booking period between the moment $t = 6$ and the final previous booking period $t = 5$.
- Finally, as shown in Fig. 2c, the user extends its booking period, and the car is used for a longer period than the first booking. Note that, in this case, we accounts for the total observed period.

We perform this interactive process to register the states of all vehicles during the whole data capture period.

---

[2] http://modo.coop/api/.

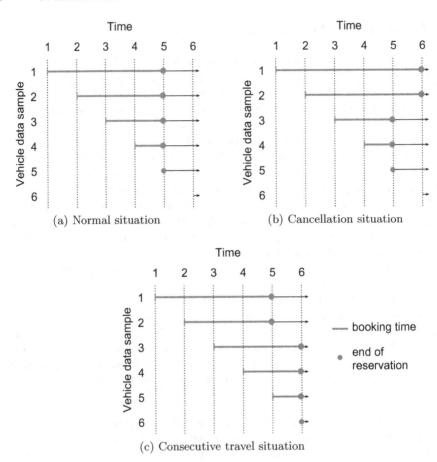

Fig. 2. Possible vehicle situations during a measurement. In (a) a normal booking and usage situation until the time t = 5; (b) a cancellation situation can be observed from the data sample 3; (c) a consecutive booking situation can be observed from sample 3.

In addition to vehicles availability information, we also collect base stations location, vehicle models, accessories and whether the vehicle is electric or hybrid. Researchers interested in the data can contact the authors.

Table 1 summarizes the volume of data collected/analyzed in this work. We have collected more than 82 k records, corresponding to 3 months of data—21 October 2017 to 21 November 2017; 01 March 2018 to 27 April 2018—, from a fleet of 592 vehicles, distributed in 471 stations, each of them with one or more cars, in the territory of Vancouver (Canada) and regions in its surroundings. We were able to analyze about 97 k booking records and more than 66 k travels.

Table 1. Summary of the data collection.

Description		Amount
# of collected records		82.275.495
# of booking records		97.865
# of travels records		66.371
# of stations		471
# of vehicles	- Common	475
	- Hybrids	114
	- Electrical	3

4 Characterizing the Car-Sharing Usage

In this section, we present the characterization about the car-sharing system service. First, we present the service temporal analysis. Then, we present the service spatial-temporal characteristics. Finally, we present service user behavior characteristics.

4.1 Service Temporal Characteristics

We now present the car-sharing service characterization. First, we present—in Fig. 3—the service daily demand pattern. Figures 3a and b present the demand pattern during labor days and Figs. 3c and d present the demand for weekends. All figures present a minute-by-minute mean value, and standard deviation, for the percentage of busy (blue curve) and reserved cars (red curve). In this case, we compute mean values and standard deviation taking into account the same one-minute period, for all days in our dataset[3].

Note that, according to Figs. 3a to d, there is a considerable difference between the number of reserved and effectively used cars, which occurs due to service cancellations. During labor days, the number of cancellations is more prominent during peak load periods, as during the middle day lunchtime and the end of the labor day. During weekends, we observe only a modal pattern, which occupies all the day.

More precisely, during labor days (Figs. 3a and b), we observe two load peaks, occurring between 11 AM–4 PM and 7 PM–8 PM. The first peak initial growth starts at 8 AM and extends until almost 6 PM. The nocturne peak may occur during after-work happy hours. For both periods, we notice a larger amount of reservation, when compared to the number of used cars, between 10 AM and 9 PM. The difference between reserved cars and actually used cars is smaller between 10 PM and 4 AM, especially noticed for the 2018 dataset. During this period, we also observe a lower demand for the service which can contribute to

---

[3] The standard deviation of idle and busy curves are the light red and gray curves, respectively.

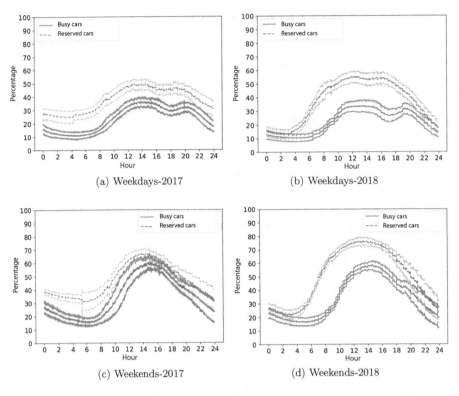

Fig. 3. Average percentage of vehicles busy/reserved at a given time for weekdays and weekends. (Color figure online)

this behavior. Finally, we notice a negligible difference between the two datasets. In this case, the only notable difference occurs by the beginning and the ending of the day, which are slightly higher for the 2017 dataset.

To analyze the characteristics of the load peaks during the working days we present in Fig. 4a and b the Empirical Cumulative Distribution Function (ECDF) of rental durations. In this case, we evaluate the load periods of a day (i.e., starts at 11 AM to 4 PM and 7 PM to 8 PM) and also, all day data. According to these figures, about 70% of vehicles rentals presents no more than 5 h of occupation. This usage time value indicates the relationship between car-sharing demand and daily work routines, suggesting that the car remains rented during the daily work times.

At weekends (Figs. 3c and d), there is a notable behavioral difference from the presented for weekdays. On weekends, there is an occupation peak between 11 AM and 6 PM, with a maximum occupancy of about 60%, near 3 PM. The reserve peak occurs between 10 AM and 6 PM, and up to 70% of available cars are booked (around 2 PM). The increased use of cars at peak times when compared to weekdays can be explained by the fact that these are times of movement for consumption in shopping and leisure centers, characteristic of weekends.

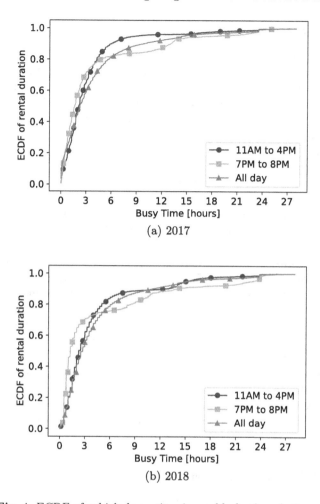

(a) 2017

(b) 2018

Fig. 4. ECDF of vehicle busy time in weekly load peak times.

In general, it is notable that the number of reservations, for the most part, is higher than that of occupancy, given that on weekends there is a greater chance of activities not having fixed time, so we conjecture that it is common for reservations to be for long periods even if they come not be fully utilized.

In summary, from Figs. 3a and b, it is observed that vehicle utilization follows usual patterns during the day, with peaks at work schedules and at night exits on weekdays, and peak in the afternoon on weekends. There was also a great variation in the number of reservations at these times, which can be explained by the great movement and availability of other means of transportation, especially in a large center such as Vancouver.

4.2   Service Spacial-Temporal Characteristics

We have also analyzed the spatial-temporal car-sharing service demand. In this sense, as shown in Fig. 5, we present heat-maps which show the hourly mean demand (occupation of vehicles) in a given base station for all data period we have analyzed. We have normalized the number of occupied vehicles to a 0–100 interval. In other words, the closer to 0, the lower the number of occupied vehicles in a given region.

Each subfigure—Fig. 5a to f—represents a 1-h interval we sampled every 4 h. We have also evaluated a distinct number of intervals, and qualitative results are similar. In this figure, we have omitted the base stations that do not present any used vehicle within the 1-h we sample.

Note that, during the first intervals (Fig. 5a and b), the demand is notable in central zones, university areas and along the rail lines, specially Expo Line and Millennium Line. In this case, we note a strong relationship between the existing public transport system and the car-sharing system demand. Indeed, during the morning period (Fig. 5c), most of the existing stations are active and we note an increasing demand for car-sharing in central regions (in special, during the period around 12 PM). During the afternoon, we note demands peaks in all regions, especially between 2 PM and 4 PM. After 6 PM, the demand on map borders starts to decrease, and the demanding focus turns back to specific points in the center, as Vancouver as well neighboring cities, universities and train stations where there are possible connections. Again, this characteristic indicates that the car-sharing users make use of public transport too. We also highlight the remarkable participation of the university public in this service. The average concentration of occupied vehicles per hour in university zones are between 50% to 100%.

4.3   User Behavior Characterization

In order to characterize the two-way car-sharing service usage, we analyze the busy and idle time of a vehicle. For this analysis, we present two cumulative distribution functions, as shown in Fig. 6, we filtered the data with less than 90 hours of duration to make the curves comparable and to avoid analyze outliers. In this figure, we plot the busy and idle vehicle time, both for common, hybrid and electric cars. We have also identified the statistical distribution that best fits the actual data. For this purpose, we tested distributions widely used in the literature: normal, lognormal, exponential, Gamma, Logistics, Beta, Uniform, Weibull and Pareto for continuous variables; Poisson, Binomial, Negative, Geometric and Hypergeometric Binomial for discrete variables. For each component of the model, the parameters of the distribution that most closely approximate the data are determined using the maximum likelihood estimation method. After defining the parameters of each component of the model, the distribution with shorter Kolmogorov-Smirnov distance (continuous distributions) or lower least square error (discrete distributions) in relation to the data was chosen. This choice is also validated with a visual assessment of the curve fitting.

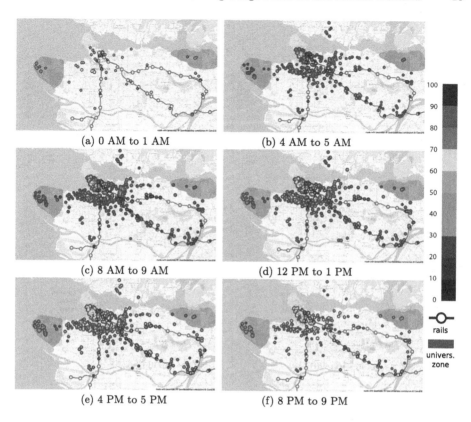

(a) 0 AM to 1 AM (b) 4 AM to 5 AM

(c) 8 AM to 9 AM (d) 12 PM to 1 PM

(e) 4 PM to 5 PM (f) 8 PM to 9 PM

Fig. 5. Percent of hourly busy vehicles in Vancouver and their environments.

Figure 6a shows the CDF of vehicle busy time, for the 2017 dataset. Note that Weibull[4] (with parameters $\alpha = 0.7074744$ and $\beta = 503.1711$) and Pareto[5] (with parameters $\alpha = 2.97408$ and $\beta = 1275.347$) distributions, present adequate adjustments to the measured data. Both distributions have similar MLEs and visually fit well into data curves. Figure 6b shows equivalent CDF, for the 2018 dataset. As occurred to the 2017 data, the Weibull and Pareto CDF distributions best adjusted the actual data, differentiating only by their parameters: Pareto - $\alpha = 0.88077$ and $\beta = 284.551$ and Weibull - $\alpha = 3.69557$ and $\beta = 812.152$. The changes in the parameters can be attributed to seasonal changes, which will be further investigated in future work.

According to Figs. 6a and b, we note that at least half of the vehicles, independent of its type, are used for more than 3 h, demonstrating that the service is used for medium to long duration travels. In addition, common and hybrid vehicles have similar characteristics however, about 40% of the electric vehicles

---

[4] Cumulative distribution function (CDF) of the Weibull distribution: $F(x; \alpha, \beta) = 1 - e^{-(x/\beta)^\alpha}$.

[5] Cumulative distribution function (CDF) of the Pareto distribution: $F(x) = 1 - \left(\frac{\beta}{x}\right)^\alpha$.

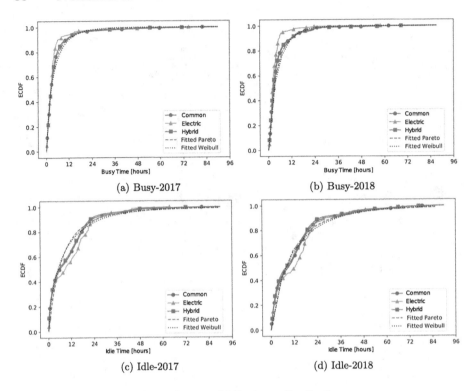

(a) Busy-2017 (b) Busy-2018

(c) Idle-2017 (d) Idle-2018

Fig. 6. Busy and idle times distribution.

remain busy 1 h less than the others. There is an indication that the lower electric vehicles busy time occurs due its intrinsic characteristics: these cars need frequent recharges, which turn them not so favorable for long-term use, but more assertive conclusions demands more electric vehicle samples.

Finally, Figs. 6c and d, presents vehicle idle periods distribution. In other words, these figures show for how long a car will be idle in its base station. It is possible to note that vehicles remain for a considerable time in their stations. Most of 60% of the cars stay for periods longer than 3 h idle. Again, common and hybrid vehicles have similar characteristics, while electric vehicles have 45% of their vehicles remaining 1 h more idle than the others. For example, at least 20% of ordinary cars get more than 18 h stationary while electric cars, in the same proportion, stay for 19 h. As we previously stated, electric vehicle intrinsic characteristic imposes hard constraints to their use, in special, forcing vehicles to be idle for long periods while they are recharging.

Weibull and Pareto distributions best fit actual data, for both datasets. In this case, both distributions favor the curves heads an tails, which comprises more than 60% of actual data. For the 2017 dataset, distribution parameters are: Weibull - $\alpha = 0.7074744$ and $\beta = 503.1711$; and Pareto - $\alpha = 2.97408$

and $\beta = 1275.347$. For the 2018 dataset, distribution parameters are: Weibull - $\alpha = 0.7940277$ and $\beta = 709.7194$) and Pareto - $\alpha = 3.34977$ and $\beta = 1944.027$).

In sum, our analysis shows that the two-way car-sharing system is mostly used for medium and long-term travels, most likely for the round trip of a work routine, or visits to cities around Vancouver. In addition, we have observed that ordinary and hybrid cars do not have notable differences in their idle and busy times, while the electric vehicle presents at least one hour difference in both cases (1 h less busy time and 1 h more idle time), probably because the need for constant recharges. Again, more assertive conclusions about electric cars demand more samples and analysis.

5   Related Work

The characterization of the usage and demands of car-sharing systems have so far been focused on one-way and free-float models. Some authors generalize one-way and two-way models as station-based models as station-based models (Becker et al. 2017; Ciari et al. 2014; Rooke et al. 2018; Boldrini et al. 2016). This paper explores the two-way model due to its specific characteristics, and the fact that this model has been little approached in the literature.

Most of the recent research (Becker et al. 2017; Ciari et al. 2014; Martin and Shaheen 2011; Boldrini et al. 2016) addressed one-way car-sharing services, revealing some important characteristics of these services. In special, these works characterize user behavior, revealing that the services are mostly used for long journeys and for shopping (Ciari et al. 2014). In most cases, the car is used by 2 or more passengers (Becker et al. 2017). These works also reveal interesting features about the fleet of electric cars: e.g., vehicles remain parked in central regions much less than in suburban regions, directly interfering with the autonomy of the vehicle (Boldrini et al. 2016). Finally, it is important to emphasize that these works indicate a close interaction between the use of the services and the use of public transport, in particular, the rail transport system (Stillwater et al. 2009).

The free-floating model clearly presents different usage patterns from the one-way model. Indeed, the free-floating vehicles are often used for shorter journeys, presenting pendular movements and a considerable number of trips to airports (Ciari et al. 2014; Becker et al. 2017; Ciociola et al. 2017). Typically, free-floating vehicles carry a single user (Becker et al. 2017). The free-floating model also presents a remarkable seasonal usage. For instance, during the mornings, central areas of the city are the main destination. During the evening, suburban areas turn the main targets (Ciociola et al. 2017).

Despite the flexibility of the free-floating model, previous works have not observed a clear difference in users' preferences between services (Ciari et al. 2014), which may be a signal that the services complement each other. Some works have identified that the services attract different user classes, exposing the fact that free-floating models and station-based models must be treated separately (Becker et al. 2017).

6   Conclusions and Future Work

In this paper, we have studied the usage pattern and service demand of a large two-way car-sharing system. Our work relies on real datasets from Modo, a car-sharing service that operates in Vancouver (Canada) and nearby regions. Our work reveals important two-way car-sharing properties, as its demand seasonality, vehicles occupation duration, travels cancellation and the waste of productivity while vehicles are idle. At a glance, our work shows that this service presents load peaks during the day. In special, during labor days, these peaks occur around the lunch time. At weekends, peaks occur during the afternoons. Most travels are rest for about 3 h and, electric vehicles stays occupied about 1 h less than other vehicles, and also present a longer maintenance period. Moreover, the car-sharing system usage presents a strong relation with public transport system, as well as with regions nearby points of interests, such as public universities and commercial centers. Finally, we believe the characterization we provide may be used as a substrate for urban centers planning.

As future work, we intend to analyze the car-sharing system regardless its type. We also plan to include additional parameters to the analysis, such as socioeconomic and environmental data.

References

Becker, H., Ciari, F., Axhausen, K.W.: Comparing car-sharing schemes in Switzerland: user groups and usage patterns. Transp. Res. Part A: Policy Pract. **97**, 17–29 (2017)

Boldrini, C., Bruno, R., Conti, M.: Characterising demand and usage patterns in a large station-based car sharing system. In: 2016 IEEE Conference on Computer Communications Workshops (INFOCOM WKSHPS), pp. 572–577. IEEE (2016)

Burkhardt, J., Millard-Ball, A.: Who is attracted to carsharing? Transp. Res. Rec.: J. Transp. Res. Board **1986**, 98–105 (2006)

Cervero, R., Tsai, Y.: City carshare in San Francisco, California: second-year travel demand and car ownership impacts. Transp. Res. Rec.: J. Transp. Res. Board **1887**, 117–127 (2004)

Ciari, F., Bock, B., Balmer, M.: Modeling station-based and free-floating carsharing demand: test case study for Berlin. Transp. Res. Rec.: J. Transp. Res. Board **2416**, 37–47 (2014)

Ciociola, A., et al.: UMAP: urban mobility analysis platform to harvest car sharing data (2017)

Cocca, M., Giordano, D., Vassio, L., Mellia, M.: Free floating electric car sharing in smart cities: data driven system dimensioning (2018)

Harms, S., Truffer, B.: The emergence of a nationwide carsharing co-operative in Switzerland. A case-study for the EC-supported rsearch project "Strategic Niche Management as a tool for transition to a sustainable transport system", EAWAG, Zürich (1998)

Herrera, J.C., et al.: Evaluation of traffic data obtained via GPS-enabled mobile phones: the mobile century field experiment. Transp. Res. Part C: Emerg. Technol. **18**(4), 568–583 (2010)

Kopp, J., Gerike, R., Axhausen, K.W., et al.: Do sharing people behave differently? An empirical evaluation of the distinctive mobility patterns of free-floating car-sharing members. Transportation **42**(3), 449–469 (2015)

Ma, S., Zheng, Y., Wolfson, O.: T-share: a large-scale dynamic taxi ridesharing service. In: 2013 IEEE 29th International Conference on Data Engineering (ICDE), pp. 410–421 (2013)

Martin, E., Shaheen, S.: The impact of carsharing on public transit and non-motorized travel: an exploration of North American carsharing survey data. Energies **4**(11), 2094–2114 (2011)

Millard-Ball, A.: Car-Sharing: Where and How It Succeeds, vol. 108. Transportation Research Board (2005)

Nourinejad, M.: Dynamic optimization models for ridesharing and carsharing. Master's thesis, University of Toronto (2014)

Rooke, F., Aquiles, V., Vieira, A.B., do Couto Teixeira, D., Almeida, J.M., Drago, I.: Caracterização de padrões de demanda e uso de um sistema de compartilhamento de veículos de duas vias. In: Simpósio Brasileiro de Redes de Computadores (SBRC), vol. 36 (2018)

Shaheen, S.A.: Mobility and the sharing economy. Transp. Policy **51**(Suppl. C), 141–142 (2016)

Shaheen, S.A., Rodier, C., Murray, G., Cohen, A., Martin, E.: Carsharing and public parking policies: assessing benefits, costs, and best practices in North America. Technical report (2010)

Stillwater, T., Mokhtarian, P., Shaheen, S.: Carsharing and the built environment: geographic information system-based study of one us operator. Transp. Res. Rec.: J. Transp. Res. Board **2110**, 27–34 (2009)

MobilityMirror: Bias-Adjusted Transportation Datasets

Luke Rodriguez[1(✉)], Babak Salimi[1], Haoyue Ping[2], Julia Stoyanovich[3],
and Bill Howe[1]

[1] University of Washington, Seattle, USA
{rodriglr,billhowe}@uw.edu, bsalimi@cs.washington.edu
[2] Drexel University, Philadelphia, USA
hp354@drexel.edu
[3] New York University, New York City, USA
stoyanovich@nyu.edu

Abstract. We describe customized synthetic datasets for publishing mobility data. Companies are providing new transportation modalities, and their data is of high value for integrative transportation research, policy enforcement, and public accountability. However, these companies are disincentivized from sharing data not only to protect the privacy of individuals (drivers and/or passengers), but also to protect their own competitive advantage. Moreover, demographic biases arising from how the services are delivered may be amplified if released data is used in other contexts.

We describe a model and algorithm for releasing origin-destination histograms that removes selected biases in the data using causality-based methods. We compute the origin-destination histogram of the original dataset then adjust the counts to remove undesirable causal relationships that can lead to discrimination or violate contractual obligations with data owners. We evaluate the utility of the algorithm on real data from a dockless bike share program in Seattle and taxi data in New York, and show that these adjusted transportation datasets can retain utility while removing bias in the underlying data.

1 Introduction

Urban transportation continues to involve new modalities including rideshare [26], bike shares [41], prediction apps for public transportation [16], and routing apps for non-motorized traffic [6]. These new services require sharing data between companies, universities, and city agencies to enforce permits, enable integrative models of demand and ridership, and ensure transparency. But releasing data publicly via open data portals is untenable in many situations: corporate data is encumbered with contractual obligations to protect competitive advantage, datasets may exhibit biases that can reinforce discrimination [18] or damage

J. Stoyanovich—This work was supported in part by NSF Grant No. 1741047.

© Springer Nature Switzerland AG 2019
J. Oliveira et al. (Eds.): BiDU 2018, CCIS 926, pp. 18–39, 2019.
https://doi.org/10.1007/978-3-030-11238-7_2

the accuracy of models trained using them [28], and all transportation data is inherently sensitive with respect to privacy [12]. To enable data sharing in these sensitive situations, we advocate releasing "algorithmically adjusted" datasets that *destroy causal relationships between certain sensitive variables* while *preserving relationships in all other cases.*

For example, early deployments of transportation services may favor wealthy neighborhoods, inadvertently discriminating along racial lines due to the historical influence of segregation [1]. Releasing data "as is" would complicate efforts to develop fair and accurate models of rider demand. For example, card swipe data for public transportation use in Seattle is biased toward employees of tech companies and other large organizations, while other neighborhoods typically use cash. This bias correlates with race and income, potentially reinforcing social inequities. Additionally, the privacy concerns of releasing this kind of data in raw form require careful attention.

Our focus in this paper is to model how these biased effects manifest in the context of transportation and how to correct for them in the context of individual privacy. We will consider three applications: ride hailing services (using synthetic data), taxi services (using public open data), and dockless bike share services (using sensitive closed data).

We focus on dockless bikeshare services as a running example. The City of Seattle began a pilot program for dockless bikes in the Summer of 2017, issuing permits for three different companies to compete in the area (Company A, B, and C). To ensure compliance with the permits, these three companies are required to share data through a third-party university service to enable integrative transportation applications while protecting privacy and ensuring equity. As part of this project, the service produces synthetic datasets intended to balance the competing interests of utility, privacy, and equity. Figure 1 shows a map of the ridership for the pilot program in Seattle and is indicative of the kind of data products generated for transparency and accountability reasons.

There are several potential *sensitive causal dependencies* in these datasets:

- Company A may be moving their bikes into particular neighborhoods to encourage commutes; this strategy could be easily copied at the cost of competitive advantage.
- Company B may be marketing to male riders through magazine ads, leading to a male bias in ridership that could be misinterpreted as demand.
- Company C may be negotiating with the city for subsidies for rides in underserved neighborhoods; they may be disallowed from publicly disclosing information about these subsidies, and therefore wish to remove the relationship between company and demographics.
- Ride hailing and taxi services allow passengers to rate and tip the drivers; gender or racial patterns in tips or ratings may encourage discrimination by drivers and should be eliminated before attempting to develop economic models of tip revenue.

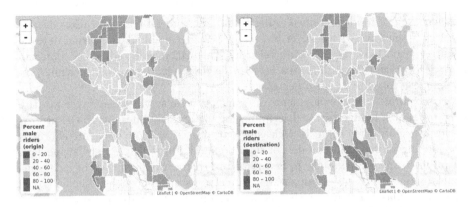

Fig. 1. Percentage of bikeshare trips in Seattle with male riders by origin and destination neighborhoods

In this paper, we develop an approach for adjusting transportation datasets to remove the effects of these sensitive causal relationships while preserving utility for classification and other analysis tasks.

Transportation data is frequently released as an *Origin-Destination* (OD) dataset: a set of location pairs representing city blocks, neighborhoods, or other spatial aggregation unit along with the traffic flow between the pair of locations. We augment OD datasets with metadata, such that each tuple represents a histogram bucket corresponding to a unique combination of attributes. For example, the bike share data includes an attribute *gender* with domain (*male, female, other*) and an attribute *company* with domain (*A, B, C*) in addition to *origin* and *destination* attributes, each with a domain of 90 neighborhoods around Seattle. A released dataset then might include the tuple (*female, A, Downtown, Ballard,* 245) indicating that there were 245 trips taken by female riders on bikes owned by company B from Downtown to Ballard during the time period covered by the dataset. These generalized OD datasets are sufficient for a variety of analytics tasks, including modeling demand, evaluating equity, estimating revenue, analyzing ridership trends, and estimating the effect on parking and motorized traffic.

Although these datasets are aggregated, they can still expose sensitive information. Individual privacy is an important concern in data sharing, but we do not focus on it here. In this work, we are interested in other forms of sensitive information encoded in the joint distribution across attributes. To remove these sensitive patterns, the data publisher specifies a causal relationship between two attributes that they wish to eliminate in the adjusted dataset, conditioned on another set of attributes Z. Then the causal repair problem is to set the mutual information between X and Y to zero, conditioned on Z. The conditional attributes Z are important to express realistically complex situations; without these attributes, degenerate solutions such as scrambling or removing the X or Y attribute altogether would be sufficient.

In our transportation context, our approach corresponds to computing a new distribution of trips over the buckets, one that preserves certain conditional joint probabilities while making other joint probabilities independent. We apply this approach to two real-world datasets of interest: the NYC taxi trip record dataset [32] and dockless bikeshare data from the city of Seattle. The NYC taxi dataset is already available; we choose to evaluate on this dataset to ensure reproducibility. The bikeshare data is legally encumbered and cannot be shared publicly.

To evaluate the efficacy of our bias-reduction approach we show that the distance between the original data and the adjusted data, as measured by multiple appropriate distance metrics, is no greater than would be expected due to sampling variance.

Since our ultimate goal is to be able to release the datasets that we produce, we also investigate how we can adapt existing techniques from the differential privacy literature to work alongside our bias reduction approach. This requires that we carefully consider the domain of the data in order to produce results that have meaningful privacy guarantees. Moving beyond the theory, we also provide results showing how privacy preservation and bias reduction interact with each other in practice.

We make the following contributions:

- We describe the bias repair problem for transportation data, which arose from collaborations with companies and city agencies interested in sharing sensitive transportation data.
- We describe a solution for removing a causal dependency (as defined by conditional mutual information) between two attributes in the context of transportation data.
- We evaluate this method on a synthetic rideshare dataset, a real taxi dataset, and a real bikeshare dataset, and demonstrate its effectiveness.
- We discuss generalizations of this approach to other domains, as we well as new potential algorithms to handle specific cases.
- We evaluate how a carefully designed privacy-preserving algorithm that can be used in conjunction with our first solution to preserve utility while formalizing privacy protections.

The rest of this paper is organized as follows: in Sect. 2 we describe related work in data sharing, causal analysis, and transportation. In Sect. 3 we present problem model and our proposed algorithm. We describe taxi and bike sharing applications in Sect. 4, and in Sect. 5 we evaluate the algorithm on real and synthetic data. We then extend these results in section Sect. 6 by presenting an algorithm for preserving privacy along with an emperical evaluation of its performance. We conclude and discuss possible extensions in Sect. 7.

2   Related Work

Recent reports on data-driven decision making underscore that fairness and equitable treatment of individuals and groups is difficult to achieve [3, 4, 30], and that transparency and accountability of algorithmic processes are indispensable but rarely enacted [5,10,36]. Our approach combines theoretical work relating causality to fairness [21] with practical tools for pre-processing data.

Recent research considers fairness, accountability and transparency properties of specific algorithms and their outputs. Dwork et al. articulated the fairness problem, emphasizing individual fairness (similar individuals receive similar outcomes), and Zemel et al. presented a method for learning fair representations based on this model that suppress discriminatory relationships while preserving other relationships [40]. Feldman et al. provided a formalization of the legal concept of disparate impact [15]. Zliobaite presented a survey of 30+ fairness measures in the literature [42]. However, these approaches are limited by the assumption that no information and no intervention methods are available for the upstream process that generated the input data [22]. Our focus is on developing a practical methodology that improves fairness for these upstream processes, specifically biased transportation data.

A common class of approaches to interrogate fairness and quantify discrimination is to use an associative (rather than a causal) relationship between a protected attribute and an outcome. One issue with these approaches is that they do not give intuitive results when the protected attribute exhibits spurious correlations with the outcome via a set of covariates. For instance, in 1973, UC Berkeley was sued for discrimination against females in graduate school admissions,when it was found that 34.6% of women were admitted in 1973 as opposed to 44.3% of men. However, it turned out that women tended to apply to departments with lower overall acceptance rates; the admission rates for men and women when conditioned on department was approximately equal [35]. The data could therefore not be considered evidence for gender-based discrimination.

The importance of causality in reasoning about discrimination is recognized in recent work. Kusner articulated the link between counterfactual reasoning and fairness [24]. Datta et al. introduce quantitative input influence measures that incorporate causality for algorithmic transparency to address correlated attributes [9]. Galhotra et al. use a causal framework to develop a software testing framework for fairness [17]. Kilbertus et al. formalize a causal framework for fairness that is closely related to ours, but do not present an implementation or experimental evaluation [21]. Nabi and Shpitser use causal pathways and counterfactuals to reason about discrimination, use causality to generalize previous proposals for fair inference, and propose an optimization problem that recomputes the joint distribution to minimize KL-divergence under bounded constraints on discrimination [31]. However, they do not provide an experimental evaluation, and do not propose an algorithm to eliminate causal relationships altogether. No prior work uses these frameworks to generate synthetic data. In our work, we focus on discrimination through total and direct effect of a sensitive attribute on an outcome. A comprehensive treatment of discrimination through causality requires reasoning about *path-specific causality* [31], which is difficult to measure in practice, and is the subject of our future work.

Prior work on publishing differentially private histograms, an intermediate step in generating synthetic data, was summarized by Meng et al. [29]. All previous approaches assume a known, fixed domain, and look to improve utility over the basic approach for numeric data proposed by Dwork et al. [14].

The first family of extensions are those that use hierarchical histogram structures. Xiao et al. propose a technique for using subcube histograms to improve accuracy in which the inputs are already binned into ranges [38]. Similarly, the concept of universal histograms helps Hay et al. [20] improve on the accuracy of Dwork's simple technique. Even so, they take as input a domain tree of unit intervals for the construction of the universal histogram. Building on the intuition that histograms depend heavily on bin choice, NoiseFirst and StructureFirst explicitly address both issues (Xu et al. [39]).

More recently there have been a few generalized techniques proposed that are adaptable to the case of histogram publication. Privelet [37] uses wavelet transforms and takes as input a frequency matrix of the counts to be approximated with noise proportional to the log of the number of such counts. Rastogi and Nath established the Fourier Perturbation Algorithm (FPA) [33], but in this framework the queries must explicitly define the domain before the algorithm is run. Building on this work, Acs et al. focus on the histogram problem and propose an extension of FPA called EFPA and a new algorithm P-HPartition as solutions [2]. While both of these methods improve on the results of FPA for histogram publication, both explicitly take a "true" histogram as input complete with defined bins. Another more general method often extended to histogram publication is DPSense [11]. The authors present this algorithm explicitly as a method for the answering the query for a vector of column counts, which in themselves encode previous assumptions about the domain.

Lu et al. consider generating synthetic data for testing untrusted systems [25], but assume a matrix structure to the data that implies a known domain. Similarly, Xiao et al. consider synthetic data release through multidimensional partitioning, but use data cubes to explicitly map to a bounded n-dimensional space. There is also work that extends these ideas to correlated attributes and graphical models, most famously the PrivBayes algorithm [7]. However, this also relies on a bounded domain in order to draw inferences about the correlation structure. The approach we adopt to account for large uncertain domain sizes is closely related to the sparse vector technique proposed by Cormode et al. [8], which explicitly models elements beyond the active domain observed in the dataset.

3   Model and Algorithm

In this section, we model the bias repair problem, provide some background on causality, and present our solution. We interpret the problem of removing bias from a dataset as eliminating a *causal dependency* between a *treatment attribute* X and an *outcome attribute* Y, assuming *sufficient covariates* $\mathbf{Z}$.

X and Y are conditionally independent given $\mathbf{Z}$ in R, written $(X \perp\!\!\!\perp Y | \mathbf{Z})$, if

$$P_R(X, Y, \mathbf{Z}) = P_R(X, \mathbf{Z}) P_R(Y | \mathbf{Z})$$

The strength of a causal link between X and Y is measured by the conditional mutual information between X and Y given $\mathbf{Z}$ [35]. It holds that $(X \perp\!\!\!\perp Y | \mathbf{Z})$ iff the conditional mutual information between X and Y given $\mathbf{Z}$ is zero, written $I(X; Y | \mathbf{Z})$. To remove bias is to enforce $(X \perp\!\!\!\perp Y | \mathbf{Z})$ or, equivalently, to set the conditional mutual information between the treatment and the outcome given the sufficient covariates to zero.

Following an example from the introduction, we can consider the effect of bike share company on gender: one company may market more aggressively to women, or their bikes may be more difficult for men to ride. This causal dependency warrants removal in various situations. For instance, the company may not want to reveal their marketing strategy, they may not want to be seen as propagating a gender bias, or a model trained on these results may be less generalizable to other cities if this bias is propagated.

Problem Statement: Bias Repair. *Given a relation R with a causal dependency $(X \not\perp\!\!\!\perp Y | \mathbf{Z})$, and given a dissimilarity measure Δ between two probability distributions, the* bias elimination problem *is to find R' such that $(X \perp\!\!\!\perp Y | \mathbf{Z})$ while minimizing $\Delta(R, R')$.*

The dissimilarity measure Δ is interpreted as between $P_R(\mathbf{A})$ and $P_{R'}(\mathbf{A})$ (e.g., KL-divergence). We will consider two different distance metrics in Sect. 5.1: Position-weighted Kendall's Tau (which is rank-sensitive) and Hellinger distance (which is not). We defer a theoretical study of this optimization problem to our ongoing and future work, though we point out a connection to the problem of low-rank matrix approximation [27]. In this paper, we propose an algorithm that directly enforces the independence condition, then show experimentally that the effect on distance is small.

3.1   Background on Causality

We now briefly review causal inference, which forms the basis of our repair algorithm. The goal of causal inference is to estimate the effect of a *treatment* attribute X on an *outcome* attribute Y while accounting for the effects of covariate attributes $\mathbf{Z}$. We compute a *potential outcome* $Y(x)$ [34], which represents the outcome if, in a hypothetical intervention, the value of X were set to value x. The *causal effect of* X *on* Y is the expected value of the difference in the potential outcomes for two different values of X: $E[Y(x_1) - Y(x_0)]$, called the *average treatment effect (ATE)*.

ATE can be computed if we can assume that a) missing attributes can be treated as having values that are effectively assigned at random (unconfoundedness/ignorability), and that b) it is possible to observe both positive and negative examples of X in a relevant subset of the data (overlap). These two conditions can be formalized as assuming a subset of attributes $\mathbf{Z} \subseteq \mathbf{A}$ is available such that:

$\forall \mathbf{z} \in Dom(\mathbf{Z}),$

$$Y(x_0), Y(x_1) \perp\!\!\!\perp X \mid \mathbf{Z} = \mathbf{z} \qquad \text{(Unconfoundedness)}$$
$$0 < \Pr(X = x_1 \mid \mathbf{Z} = \mathbf{z}) < 1 \qquad \text{(Overlap)}$$

If these conditions are met, ATE can be computed as follows:

$$\text{ATE} = \sum_{\mathbf{z} \in Dom(\mathbf{Z})} \left(\mathbb{E}[Y|X = x_1, \mathbf{Z} = \mathbf{z}] - \mathbb{E}[Y|X = x_0, \mathbf{Z} = \mathbf{z}] \right) \Pr(\mathbf{Z} = \mathbf{z}) \qquad (1)$$

where $Dom(\mathbf{Z})$ is the domain of the attributes $\mathbf{Z}$.

From this expression, it can be shown that the ATE of X on Y is zero iff $I(X;Y|\mathbf{Z}) = 0$. Therefore, we can use the conditional mutual information $I(X;Y|\mathbf{Z})$ to quantify the strength of a causal link between X and Y given $\mathbf{Z}$.

ATE quantifies the *total* effect of X on Y, which can be separated into direct effects and indirect effects (those that are mediated through other attributes). In this paper, we ignore this distinction, and leave generalizing the method to account for this distinction to future work.

3.2   Algorithm

We propose a simple algorithm to compute an approximate solution to our problem. The algorithm is based on the intuition that $(X \perp\!\!\!\perp Y|\mathbf{Z})$ holds in R' iff the joint probability distribution $\Pr_{R'}(\mathbf{A})$ admits the following factorization, based on the chain rule:

$$P_{R'}(\mathbf{A}) = P_{R'}(X\mathbf{Z})P_{R'}(Y|\mathbf{Z})P_{R'}(\mathbf{U}|XY\mathbf{Z}) \qquad (2)$$

where $\mathbf{U} = A - XY\mathbf{Z}$.

This factorization will form the basis of our algorithm, but there is a complication: We want to restrict R' to include only the active domain of R rather than the full domain. The reason is that transportation datasets are typically sparse; there are many combinations of attributes that do not correspond to any traffic (e.g., bike rides from the far North of the city to the far South). We assume R is a bag; it may contain duplicates. For example, there may be multiple trips with the same origin, destination, and demographic information. Under this semantics, we express our algorithm in terms of *contingency tables*.

A contingency table over a set of attributes $\mathbf{X} \subseteq \mathbf{A}$, written $\mathcal{C}_R^{\mathbf{X}}$, is simply the count of the number of tuples for each unique value of $\mathbf{x} \in Dom(\mathbf{X})$. That is, $\mathcal{C}_R^{\mathbf{X}}$ corresponds to the result of the query select $\mathbf{X}$, count(*) from R group by $\mathbf{X}$. More formally, a contingency table over $\mathbf{X} \subseteq \mathbf{A}$ is a function $Dom(\mathbf{X}) \to \mathbb{N}$

$$\mathcal{C}_R^{\mathbf{X}}(\mathbf{x}) = \sum_{t \in R} \mathbb{1}[t[\mathbf{X}] = \mathbf{x}]$$

$t[\mathbf{X}]$ represents the tuple t projected to the attributes $\mathbf{X}$, and $\mathbb{1}$ is the indicator function for the condition $t[\mathbf{X}] = \mathbf{x}$. The contingency table over all attributes in R is an alternative representation for the bag R itself: Given $\mathcal{C}_R^{\mathbf{A}}$, we can recover R by iterating over $Dom(\mathbf{A})$. In practice, this step is not necessary, as $\mathcal{C}$ is implemented as a k-dimensional array.

Using contingency tables, we can compute a new joint probability distribution over $\mathbf{A}$ as

$$P_R(\mathbf{A} = a) = \frac{\mathcal{C}_R^{\mathbf{A}}(a)}{|R|}$$

Algorithm 1 uses these ideas to construct the desired relation R' from the marginal frequencies of R, enforcing Eq. 2 by construction. It can be shown that the KL-divergence between $P_R(\mathbf{A})$ and $P_{R'}(\mathbf{A})$ is bounded by $I(X;Y|\mathbf{Z})$. That is, the divergence of R' from R depends on the strength of the causal dependency between X and Y. If the causal dependency is weak, Algorithm 1 will have no significant effect on the dataset. We will evaluate the effects experimentally in Sect. 5.

Algorithm 1. Enforcing Conditional Independence

    Input: An instance R with $\mathbf{A} = XY\mathbf{Z}\mathbf{U}$ in which $(X \not\!\perp Y|\mathbf{Z})$

    Output: An instance R' in which $(X \perp\!\!\!\perp Y|\mathbf{Z})$

1 **for** $xy\mathbf{z}\mathbf{u} \in R$ **do**

2      $numerator \leftarrow \mathcal{C}_R^{X\mathbf{Z}}(x\mathbf{z})\mathcal{C}_R^{Y\mathbf{Z}}(y\mathbf{z})\mathcal{C}_R^{XY\mathbf{Z}\mathbf{U}}(xy\mathbf{z}\mathbf{u})$

3      $denominator \leftarrow |R|\mathcal{C}_R^{\mathbf{Z}}(\mathbf{z})\mathcal{C}_R^{XY\mathbf{Z}}(xy\mathbf{z})$

4      $\mathcal{C}_{R'}^{\mathbf{A}}(xy\mathbf{z}\mathbf{u}) \leftarrow \mathbf{Round}(\frac{numerator}{denominator})$

5 **return** R' associated with $\mathcal{C}_{R'}^{\mathbf{A}}$

---

4   Applications and Datasets

In this section we describe two real datasets to which we apply our methodology and an overview of how both datasets were processed for use in our evaluation.

NYC Taxi Data. The NYC taxi trip record dataset released by the Taxi & Limousine Commission (TLC) [32] contains trips for 13,260 taxi drivers during January 2013, with pick-up and drop-off location as (lat,lon) coordinates and other information including trip distance and tip amount. We used this particular release of the data because medallion numbers were no longer made available after this release. We first removed transportation records with missing values, such as records with unknown pick-up or drop-off locations or missing tip amount. We then categorized trip distance into low, medium, and high, with about 1/3 of the trips falling into each category. Tip amount was categorized into low and high, with high tip corresponding to at least 20% of the fare amount. Note that the original dataset has tip amount information only for rides that were paid by a credit card, and so we only consider these trips in the paper. Lastly, drivers were categorized into low, medium, and high frequency drivers.

Table 1 shows an example of the data after aggregation, with the count of each instance represented in the `count` column.

Table 1. Processed NYC taxi data

o_lon	o_lat	d_lon	d_lat	Pickup	Dist	Tip	Freq	Count
−74.0	40.7	−74.0	40.7	Night	Med	High	Low	6074
−74.0	40.7	−74.0	40.7	Night	Med	Low	Low	2844
−73.9	40.7	−73.9	40.7	Day	Low	High	Med	16
−73.9	40.7	−73.9	40.7	Morn	Low	High	Low	14
−73.9	40.7	−74.0	40.7	Morn	Low	High	High	3

Dockless Bikeshare. The bike data includes rides from 197,049 distinct riders between June 2017 and May 2018 across three different companies. Each rider is identified via a unique rider id for each company, and the start and end location of each trip is projected to one of 94 neighborhoods in the Seattle area. Trip information is joined with rider information from survey responses, indicating their gender and whether or not they use a helmet.

Data Processing and Aggregation. We pre-processed both datasets to make them compatible with our approach. First, the time in both is precise up to the second. Since our model assumes categorical attributes, we map time to four buckets: morning (5am–9am), day (9am–3pm), evening (3pm–7pm), and night (7pm–5am). Additionally, each individual driver/rider was classified into one of three categories by the number of trips they made, as recorded in the dataset. The top 1/3, who made the most trips, are designated `heavy`, the bottom 1/3 are designated `light`, and the rest are designated `medium`.

5 Experiments

In this section, we first outline our evaluation metrics, and then present experiments to consider whether the error introduced by our bias-repair method is comparable to the error introduced by natural variation. Recall that we wish to remove the causal dependency between X and Y. If there is no correlation between these attributes, then the repair process will not change the weights significantly. If there is a strong correlation, then the process will force the mutual information to zero while preserving the distribution of the other attributes.

We consider three situations: synthetic data simulating extreme situations (Sect. 5.2), real datasets representing bike and taxi data (Sect. 5.3), and the same real bike and taxi data, but aggregated post hoc to simple origin-destination pairs (Sect. 5.4). The experiments in each of these situations can be summarized by the choice of treatment (X), outcome (Y) and covariate (Z) attributes, $X \rightarrow Y|Z$, as follows:

1. Synthetic: $gender \rightarrow rating | \{origin, destination\}$
2. Bike: $company \rightarrow gender | \{start\_nhood, end\_nhood, time\_of\_day, helmet\}$
3. Taxi: $distance \rightarrow tip | \{orig\_lon, orig\_lat, dest\_lon, dest\_lat\}$

5.1   Evaluation Metrics

Our goal is to remove the effect of the given relationship without destroying the utility of the resulting dataset. The proposed method would not be viable if it altered the distribution of traffic "too much." To define "too much," we a) compute the distance between the original dataset and the adjusted dataset, and b) compare this distance with the distances associated with a set of bootstrap samples of the original dataset. If the distance with the adjusted dataset falls within the distribution of the bootstrap samples, we conclude that the adjustment is small enough to still produce a useful dataset.

To compute distances, we consider two different metrics: one that is rank-sensitive, and one that is not. To measure rank-sensitive distance, we sort the buckets by trip count in descending order before and after the repair. We then use position-weighted Kendall's tau [23] to compare the two resulting rankings. Kendall's tau counts the number of pair-wise position swaps between a ground truth ranking and an experimental ranking. Position-weighted Kendall's tau incorporates a weighting function, usually to assign more importance to swaps that happen closer to the beginning of the ranked list.[1] This measure is appropriate in our domain, because a) transportation analysts and engineers are primarily interested in the conditions associated with the heaviest traffic flows, and b) transportation datasets are inherently sparse.

The weighting function we consider is harmonic: Given position i in a ranking, the weight is $\frac{1}{i}$. We also considered an exponential weighting function, since traffic patterns tend to follow an exponential distribution, but that weighting function was potentially too generous to our method: The first few positions were all that mattered.

To measure distance independently of rank and position, we use Hellinger distance. This measure is an f-divergence closely related to the Bhattacharyya distance that obeys the triangle inequality, and is defined as follows: Let p, q be two probability distributions over the same set of attributes $\mathbf{X}$, and define the Bhattacharyya Coefficient $BC(p, q)$ to be $\sum_{x \in \mathbf{X}} \sqrt{p(x)q(x)}$. Then the Hellinger distance is $H(p, q) = \sqrt{1 - BC(p, q)}$.

Table 2 presents results for both position-weighted Kendall's tau (PWKT) and Hellinger distance in each of our experiments. The experimental result for Algorithm 1 is in bold, and the other columns summarize the distribution of the bootstrap samples. Figure 2 visualizes these results. Each experiment is represented by three bars. The light bar on the left shows the distribution of distances

---

[1] Many methods for comparing ranked lists have been proposed. We opt for a measure in which identity of the items being ranked (histogram buckets) is deemed important. This is in contrast to typical IR measures such as NDCG or MAP, where item identity is disregarded, and only item quality or relevance scores are retained.

Table 2. Results of evaluation metrics across all experiments

PWKT	Synth. uncorrelated	Synth. correlated	Bike all	Bike grouped	Taxi all	Taxi grouped
2.5%	1.47	1.35	1.49	1.37	0.84	0.27
Mean	2.93	2.44	2.34	2.88	1.39	0.81
97.5%	4.39	3.54	3.18	4.39	1.93	1.36
Result	**0.159**	**3.18**	**1.53**	**1.21**	**1.37**	**0.40**
Hellinger	Synth. uncorrelated	Synth. correlated	Bike all	Bike grouped	Taxi all	Taxi grouped
2.5%	0.075	0.072	0.084	0.14	0.029	0.044
Mean	0.076	0.073	0.085	0.15	0.030	0.051
97.5%	0.076	0.074	0.086	0.15	0.030	0.047
Result	**0.00079**	**0.42**	**0.15**	**0.042**	**0.024**	**0.0020**

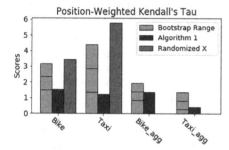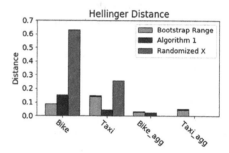

Fig. 2. Expected bootstrap variation (left), experiment outcome (center), and baseline comparison (right) for each of the four experiments on real data. Lines in the bar denote the 2.5%, mean, and 97.5% values for the distribution.

from the bootstrap procedure: the top of the bar represents the 97.5 percentile, the next line represents the mean, and lowest line represents the 2.5 percentile. We visualize the distribution as a bar to emphasize that the measure is a distance, such that a lower bar is always better. The dark bar in the center is the experimental result. The final bar on the right represents a baseline test of assigning every trip a random X value as a strategy of enforcing $I(X; Y) = 0$.

Overall, we can see that the error introduced by our algorithm is usually significantly less than the error one can expect from sampling, suggesting that the method is viable for correcting bias while retaining utility.

The expected variation is clearly visible for the case of PWKT, but for the Hellinger distance it is small compared to the magnitude of the metric, and is nearly impossible to distinguish precisely. The *Correlated* and *Bike* columns for the Hellinger distance stand out as significant outliers.

5.2 Synthetic Ride Hailing Data

For the synthetic experiments, the task is to remove the causal influence of gender on rating, simulating the situation where a data publisher does not want to unintentionally encourage discrimination [18].

To generate the synthetic data, we use neighborhood-level trip data from the dockless bikeshare to simulate a realistic distribution of traffic among neighborhoods. Then, we assign each individual trip a gender at random from $\{m, f, o\}$ representing male, female, or other. The *no correlation* experiment assigns ratings according to a pre-defined distribution independent of gender, while the *gender correlated* experiment uses three different distributions, one for each gender value, to simulate a strong correlation. In both of these experiments, our simulated repair is to remove the effect of gender on rating, conditional on the origin and destination neighborhoods.

We expect that the uncorrelated case should have minimal effect on the data, since there is no causal dependency to eliminate. For the strongly correlated case, we expect the error to be significant.

Comparing the synthetic data experiments in Table 2, we see that there is a change in the order of magnitude of the effect when the repair is acting on a relationship with a strong underlying correlation. When applied to synthetic data with no correlation structure, we find that values of both position-weighted Kendall's tau and Hellinger distance fall well below the range of error introduced by bootstrap sampling. However, in the correlated case when there was in fact a strong relationship, position-weighted Kendall's tau jumped to the upper extreme of the bootstrap range, and Hellinger distance far exceeded this range. This result indicates that the repair is causing a more drastic change in the gender correlated case than in the case with no correlation, as we would expect.

Position-weighted Kendall's tau still falls within the bootstrap variation for the correlated case, which can be explained by the fact that certain neighborhood origin-destination pairs carry a disproportionate amount of the traffic in the dataset, so this relationship is preserved. The full magnitude of the change is better observed through the Hellinger distance in Table 2, which grows an order of magnitude beyond the bootstrap variance in the gender correlated experiment.

5.3 Real-World Bike and Taxi Datasets

In the bike experiment we remove the influence of company on gender using the dockless bikeshare data described in Sect. 4. In this experiment, we are considering the situation where companies are releasing data to support traffic research, but do not want to expose any latent gender bias that may be attributable more to marketing efforts than to sexism. The relationship between company and gender is conditional on origin, destination, and whether or not the rider uses a helmet. In other words, *only the effect of company should be removed, not the overall pattern of gender on ridership.*

In the taxi experiment we investigate the effect of a repair on the taxi data from Sect. 4, in which we remove the influence of distance on tip amount, conditional on origin and destination. The situation we consider is a behavioral economic analysis of tipping patterns, but we want to completely remove the influence of distance. Simply normalizing by distance is not enough, as the joint distribution between, say, time of day, distance, and tip amount can be complex. Moreover, certain neighborhood origins and destinations may generate higher tips or lower tips in ways that interact with distance traveled. For example, long east-west trips at certain times may be relatively short, but generate higher tips.

In both cases, we see that the calculated position-weighted Kendall's tau and Hellinger distance in Table 2 fall close to the expected variation from bootstrap samples, with the exception of the Hellinger distance for the bike share data, which is about twice this baseline. This anomaly helped us discover a data ingest error upstream from our algorithm: gender information was only properly included for one company, while the other two had two different default values. As a result, there was an unrealistically high correlation between company and gender. The order of values was still largely preserved by Algorithm 1, as seen in Fig. 2, since there are significantly more trips from one company than from the others, but the structural change results in a high Hellinger distance. Taken along with the taxi data, this reaffirms that Algorithm 1 behaves as expected: it induces larger changes when there is a high degree of correlation in the relationship chosen for treatment.

5.4   Aggregated Origin-Destination Data

In our experiments so far we considered all possible fine-grained buckets in the dataset. For example, the trip count associated with {UDistrict, Downtown, Female, Helmet, Morning} appears as a bucket. We also consider a coarser aggregated view of this data, grouping buckets by origin and destination and aggregating over gender, helmet, and time. The motivation is that, in many situations, only origin-destination counts are important, and also that our method may unfairly benefit on a fine-grained dataset: if we preserve the distribution of the top few origin-destination pairs, we will also preserve the distribution of a large number of finer-grained buckets that divide these origin-destination pairs by gender, helmet and time. We run the same experiments and metrics as before, but this time grouping by origin and destination.

When aggregating as described, we see in Fig. 2 that the baseline (right column) for each of these experiments has a value of 0. This is because origin and destination were not included in X or Y, and any repair that only takes into account the relationship between X and Y does not impact the other direct relationships in the dataset. For the results of Algorithm 1 (center), the Hellinger distance falls below the expected variation for both datasets, while the position-weighted Kendall's tau falls in the bottom half of the expected range of variation. We therefore conclude that Algorithm 1 preserves both order and structure of real aggregated data at least as well as a bootstrapped sample, given these particular correlation structures.

6 Interactions with Privacy

In this section we explore how approaches in differential privacy interact with our bias-reduction algorithm. Our goal is to create an algorithm that interfaces easily with Algorithm 1 while also accounting for the fact that transportation datasets are often very sparse in the sense that there are often many combinations of attributes that do not appear in the real data, even while others are frequent (e.g. bus rides taken between neighborhoods that are not connected along a bus route). Accounting for sparsity is a key challenge, since common privacy techniques applied to large, sparse domains can destroy utility [8].

Fundamentally, aggregation techniques to ensure privacy are insufficient, especially in a transportation context. Montoye et al. showed that just four points of trajectory data are sufficient to uniquely identify most users [12]. Therefore, differential privacy techniques, where noise is added to prevent inference about individuals based on the principle of indistinguishability [13] are preferred.

Preliminaries. Two database instances D_1 and D_2 are *neighbors* if they differ by exactly one row. Next, we imagine some randomized function q and a set of results $S \subseteq Range(q)$.

Definition 6.1 (Differential Privacy). *A randomized function q is ϵ-differentially private if for all neighboring database instances D_1 and D_2 and all $S \subseteq Range(q)$,*

$$Pr[q(D_1) \in S] \leq e^\epsilon \times Pr[q(D_2) \in S] \tag{3}$$

where the probability is taken over the randomness of q.

Definition 6.1 guarantees that the two neighboring database instances are indistinguishable to within a factor of e^ϵ when presented with a result of the randomized query q. This mechanism q addresses concerns that a participant might have about leakage of their information through the results of q. Even if the participant decided not to include their data in the dataset, no outputs of q would become significantly more or less likely.

Real datasets like those in transportation use cases often run into the issue of sparse domains. For example, in the dockless bikeshare data the combination of all possible attribute values yields over 2.5 million possible distinct bins. Since many trips fall into the same bins, even with orders of magnitude more trips than bins we would expect to have a significant number of bins with no trips. By extension, this means that we need to be careful about how we reason about the bins outside of the active domain but still admissible as part of the global domain. We formalize this well-known problem as follows, and use the formalization to provide a proof that there is no way to avoid including bins from outside the active domain:

Lemma 6.1. *Given a differentially private mechanism q with respect to a set of database instances $\mathcal{D}$ of schema $R(\mathcal{A})$, if $P[q(D_i) = r] = 0$ for some database instance D_i and a result r, then $P[q(D) = r] = 0$ for all $D \in \mathcal{D}$.*

Proof. Let $P[q(D_i) = r] = 0$ for some database instance D_i and result r. For any database instance $D_j \in \mathcal{D}$, define a sequence of database instances $S(D_j, D_i) = \{D_1, D_2, \ldots, D_n\}$ such that $D_1 = D_i$, $D_n = D_j$, and D_k and D_{k+1} are neighbors for all $0 \leq k < n$. [2] It follows that

$$P[q(D_j) = r] \leq e^{\epsilon} \times P[q(D_2) = r]$$
$$\leq e^{2\epsilon} \times P[q(D_3) = r]$$
$$\vdots$$
$$\leq e^{(n-1)\epsilon} \times P[q(D_i) = r]$$
$$\leq 0$$

Since the probability cannot be less than 0, we find that $P[q(D) = r] = 0$ for all $D \in \mathcal{D}$.   □

Theorem 6.2. *There does not exist a differentially private mechanism that only returns elements from the active domain.*

The intuition for Theorem 6.2 is that every element in the global domain must have a non-zero chance of being included in the result, or else Lemma 6.1 is violated.

Proof. Let $\mathcal{D}$ be the set of possible database instances. Let c be a function such that $c(D)$ returns the active domain of D. Let c^* be some differentially private mechanism that returns a value set $c^*(D) \subseteq c(D)$. This means that c^* can be probabilistic, but only the value sets in the powerset of $c(D)$ have non-zero probabilities. Next let us pick some $D_i \in \mathcal{D}$. Then $P[q(D_i) = r] = 0$ for all $r \not\subseteq c(D_i)$, and additionally by Lemma 6.1 that $P[q(D) = r] = 0$ for all $D \in \mathcal{D}$. This holds true for all $c(D)$, and thus the possible range R of $c^*(D)$ is constrained by $R = \cap_{D \in \mathcal{D}} c(D) = \emptyset$. Therefore, our mechanism c^* must be the trivial one that returns the empty set.   □

Theorem 6.2 depends crucially on the choice of function c. If we allow this function to be defined on the global domain of the database instance D, we can in fact create a differentially private mechanism. This observation is the motivation behind the approach that we take in Algorithm 2, which is closely related to the technique outlined by Cormode et al. [8]. Our approach differs in that we parameterize our algorithm by the *tolerance for including values from outside of the active domain* rather than as a fixed number of such values. This formulation is essentially a usability enhancement: users cannot necessarily provide the number of bins (or a threshold defining the number of bins) without inspecting the dataset, while a probability for including bins can be estimated globally based on the use case: A heatmap of trips may tolerate a few trips in unexpected places,

---

[2] Two datasets are neighbors if they differ in the presence or absence of a single record, following the differential privacy definition.

while an analysis of maximum trip distance or other computations that are sensitive to outliers have a lower tolerance. We also add laplacian noise rather than geometric noise [19] for simplicity.

Our algorithm is differentially private in the context of a global domain, and controls the number of bins from outside the active domain by deriving a threshold from the user-provided tolerance ρ. Reducing ρ will lead to choosing a higher threshold and therefore excluding more of the active domain, but any choice for ρ fulfills the criteria for differential privacy. In a sense, this formulation frames the question a one of data sufficiency: Given a tolerance for irrelevant bins, do you have enough data to include the most important bins in the histogram with high probability? If not, you can still release the dataset (and retain differential privacy guarantees at the expense of utility), or you can potentially go and collect more data to produce higher bin counts and improve their likelihood of being included.

Algorithm 2. Categorical Histogram Method

Input: Difference in sizes of global and active domains n, tolerance for values outside the active domain ρ, privacy budget ϵ, and the true histogram (C, S) where C is a vector of categories in the active domain and S is a vector of true frequencies for each corresponding category in C.

Output: Differentially private histogram

1 $\tau \leftarrow -\frac{\ln(2(1-\rho^{\frac{1}{n}}))}{\epsilon}$

2 $i \leftarrow 0$

3 **while** $i < |C|$ **do**

4      $s_i \leftarrow$ LaplaceDistribution$(s_i, \frac{1}{\epsilon})$

5   **if** $s_i < \tau$ **then**

6   Remove(c_i, s_i)

7      $i \leftarrow i + 1$

8 $k \leftarrow$ BinomialDistribution$(n, \frac{1}{2}e^{-\epsilon\tau})$

9 $j \leftarrow 0$

10 **while** $j < k$ **do**

11   Append$(C, \text{GetCategoryFromDomain}())$

12   Append$(S, \tau+ \text{ExponentialDistribution}(\frac{1}{\epsilon}))$

13      $j \leftarrow j + 1$

14 **return** C, S

To see how this works in practice, we run Algorithms 1 with 2 for various values of ρ and ϵ. We can then compare these results to both the output of Algorithm 1. A key finding is that for these datasets our algorithms can be applied in either order: privacy-first, or bias-first. In other words we can compose the bias reduction and noise injection steps in either order. However, the privacy guarantee is subtly different between these two cases: If we remove bias first, the distribution is a sample from the "fair" world where biases have been

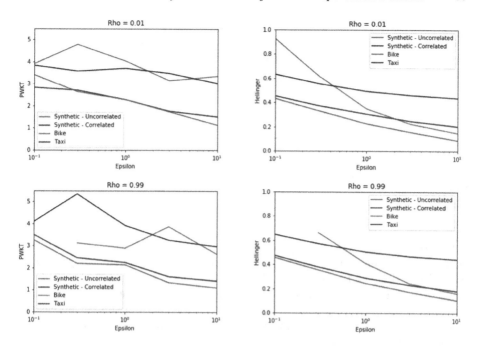

Fig. 3. Distance metrics of results of composing Algorithms 1 and 2 over all four datasets for varying values of ρ and ϵ.

removed, which is different than the "real" world. Specifically, a record in the bias-corrected dataset does not necessarily correspond to any real individual. Therefore, we present results from the privacy-first approach in order to make the privacy guarantee easier to interpret.

Figure 3 shows the results over the same four datasets and attributes that we investigated in Sect. 5 for ease of direct comparison. For the same two evaluation metrics as in the previous section we show a plot of the metric vs ϵ on a logarithmic scale for a large and small value of ρ, where each data point used comes from the average over 10 trials in order to account for the inherent randomness of the algorithm.

For some very conservative parameter combinations, the derived threshold was higher than all buckets, leading to an empty result. For example, $\epsilon = 0.1$ and $\rho = 0.99$ in the Taxi dataset implies a minimum bin count of 245. This threshold exceeded the support of any single bucket in the histogram, and our Algorithm 2 returned no data for which to calculate distances. Empty results like these demonstrate how Algorithm 2 yields differentially private results no matter what the input parameters are. In practice, the data owner could potentially choose to postpone data release until more data is collected, to increase the buckets' support and eventually produce non-empty output.

Next, we observe that the Hellinger distances (at right) are higher than those without noise, but converge to the noiseless values as the privacy budget ϵ is

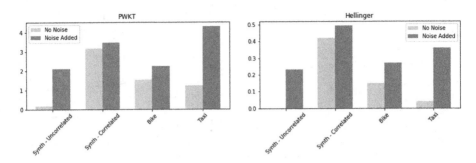

Fig. 4. Comparison of the no noise case from Sect. 5 (left) and noised results (right) for fixed $\rho = 0.5$ and $\epsilon = 1$.

increased. The rate at which they converge is directly related to the size of the global domain. Since the two synthetic datasets were created using the same set of possible origin and destination neighborhoods as the bike data, the size of the global domains is similar across all three datasets. On the other hand, the taxi data is sampled from a much smaller global domain. Therefore, increasing ϵ from 0.1 to 1 reduces the Hellinger distance by about 0.1 for the bike and synthetic datasets, but 0.5 for the taxi dataset. For low values of ϵ, the taxi results are dominated by bins included from the global domain, yielding very high distances.

However, PWKT (at left) behaves a bit less smoothly. There does still seem to be an overall decreasing trend as ϵ increases as we would expect, but this does not happen monotonically. The spikes that occur along the way come as a result of the linear decrease of the threshold chosen by 2 that comes with increasing ϵ. In particular, since PWKT is rank-sensitive, the distance it calculates depends on both the number of ranked buckets as well as their order. Thus having a high threshold that yields only a few buckets for analysis will tend to result in a smaller distance value, even if the buckets that were preserved are not preserved in order. This causes an inflection point at a different ϵ value for each value of ρ, as eventually the distance-reducing effect of more accurate ordering counteracts the distance-increasing effect of including more buckets in the distance calculation. Domain size and these inflection points also help explain why the results from the taxi data seem to fail to converge to the no-noise case towards $\epsilon = 10$, even as the other three datasets do so.

Figure 4 compares the results of composing Algorithms 1 and 2 to the no noise case presented in Sect. 5 for a reasonable choice of ρ and ϵ. Adding noise increases the distances across the board, and especially to the taxi data whose support is smaller relative to the global domain. However, we also observe that the increase in distance is larger when the distance due to bias correction was smaller. This result implies that the randomization induced by Algorithm 2 overlaps with the bias reduction of Algorithm 1. The relationship between these algorithms hints at a more fundamental relationship between the processes of bias reduction and privacy preservation, which is a potential direction for future work.

7 Conclusions and Future Work

Data sharing is emerging as a critical bottleneck in urban and social computing. While risks associated with privacy have been well-studied, data owners and data publishers must also be selective about the patterns they reveal in shared data. Biases in the underlying data can be reinforced and amplified when used to train models, leading to not only poor quality results but also potentially illegal discrimination against protected groups, causing a breach of trust between government and companies.

In this paper, we have considered the bias-correction problem—an important pre-processing step in releasing data that is orthogonal to privacy.

We interpret the need to repair unintended or unrepresentative relationships between variables prior to data release as related to causal inference: the conditional mutual information between two variables is a measure of the strength of the relationship. We propose an algorithm that interprets the frequencies of trip events as a probability distribution, then manipulates this distribution to eliminate the unwanted causal relationship while preserving the other relationships.

We show that this procedure produces expected behavior for synthetic datasets representing extreme cases, and has only a modest impact in real datasets: the distance between the original data and the adjusted data falls within the bounds of natural variation of the original data itself. Additionally, we present a method to adjust our approach to meet the standards of differential privacy - a crucial step for the adjustment and release of any real-world dataset.

Going forward, we aim to generalize this approach to other domains, distinguish between direct and indirect causal effects, and explore new algorithms that can better balance the tradeoff between utility and causal relationships. We also hope to investigate the ways that this approach interacts with privacy-preserving methods in more detail, including direct overlap that could be leveraged to decrease the amount of noise added as well as formalizing what it might mean to preserve privacy after the bias reduction step. Our broader vision is to develop a new kind of open data system that can spur data science research by generating safe and useful synthetic datasets on demand for specific scenarios, using real data as input.

References

1. Amazon doesn't consider the race of its customers. should it? Bloomberg (2016)
2. Acs, G., Castelluccia, C., Chen, R.: Differentially private histogram publishing through lossy compression. In: 2012 IEEE 12th International Conference on Data Mining (ICDM), pp. 1–10. IEEE (2012)
3. Angwin, J., Larson, J., Mattu, S., Kirchner, L.: Machine bias: risk assessments in criminal sentencing. ProPublica, 23 May 2016
4. Barocas, S., Selbst, A.: Big data's disparate impact. Calif. Law Rev. **104**, 671 (2016)
5. Brauneis, R., Goodman, E.P.: Algorithmic transparency for the smart city. Yale J. Law Technol., forthcoming

6. Brock, A.M., et al.: SIG: making maps accessible and putting accessibility in maps. In: Extended Abstracts of the 2018 CHI Conference on Human Factors in Computing Systems, p. SIG03. ACM (2018)

7. Chen, R., Fung, B.C., Yu, P.S., Desai, B.C.: Correlated network data publication via differential privacy. VLDB J. **23**(4), 653–676 (2014)

8. Cormode, G., Procopiuc, M., Srivastava, D., Tran, T.T.: Differentially private publication of sparse data. arXiv preprint arXiv:1103.0825 (2011)

9. Datta, A., Sen, S., Zick, Y.: Algorithmic transparency via quantitative input influence: theory and experiments with learning systems. In: IEEE SP, pp. 598–617 (2016)

10. Datta, A., Tschantz, M.C., Datta, A.: Automated experiments on ad privacy settings. PoPETs **2015**(1), 92–112 (2015)

11. Day, W.-Y., Li, N.: Differentially private publishing of high-dimensional data using sensitivity control. In Proceedings of the 10th ACM Symposium on Information, Computer and Communications Security, pp. 451–462. ACM (2015)

12. de Montjoye, Y.-A., Hidalgo, C.A., Verleysen, M., Blondel, V.D.: Unique in the crowd: the privacy bounds of human mobility. Sci. Rep. **3**, 1376 (2013)

13. Dwork, C.: Differential privacy. In: Bugliesi, M., Preneel, B., Sassone, V., Wegener, I. (eds.) ICALP 2006. LNCS, vol. 4052, pp. 1–12. Springer, Heidelberg (2006). https://doi.org/10.1007/11787006_1

14. Dwork, C., McSherry, F., Nissim, K., Smith, A.: Calibrating noise to sensitivity in private data analysis. In: Halevi, S., Rabin, T. (eds.) TCC 2006. LNCS, vol. 3876, pp. 265–284. Springer, Heidelberg (2006). https://doi.org/10.1007/11681878_14

15. Feldman, M., Friedler, S.A., Moeller, J., Scheidegger, C., Venkatasubramanian, S.: Certifying and removing disparate impact. In: Proceedings of the 21th ACM SIGKDD International Conference on Knowledge Discovery and Data Mining, KDD 2015, pp. 259–268. ACM, New York (2015)

16. Ferris, B., Watkins, K., Borning, A.: OneBusAway: results from providing real-time arrival information for public transit. In: Proceedings of the SIGCHI Conference on Human Factors in Computing Systems, pp. 1807–1816. ACM (2010)

17. Galhotra, S., Brun, Y., Meliou, A.: Fairness testing: testing software for discrimination. In: Proceedings of the 2017 11th Joint Meeting on Foundations of Software Engineering, ESEC/FSE 2017, Paderborn, Germany, 4–8 September 2017, pp. 498–510 (2017)

18. Ge, Y., Knittel, C.R., MacKenzie, D., Zoepf, S.: Racial and gender discrimination in transportation network companies. Working Paper 22776, National Bureau of Economic Research, October 2016

19. Ghosh, A., Roughgarden, T., Sundararajan, M.: Universally utility-maximizing privacy mechanisms. SIAM J. Comput. **41**(6), 1673–1693 (2012)

20. Hay, M., Rastogi, V., Miklau, G., Suciu, D.: Boosting the accuracy of differentially private histograms through consistency. Proc. VLDB Endow. **3**(1–2), 1021–1032 (2010)

21. Kilbertus, N., Carulla, M.R., Parascandolo, G., Hardt, M., Janzing, D., Schölkopf, B.: Avoiding discrimination through causal reasoning. In: Advances in Neural Information Processing Systems, pp. 656–666 (2017)

22. Kirkpatrick, K.: It's not the algorithm, it's the data. Commun. ACM **60**(2), 21–23 (2017)

23. Kumar, R., Vassilvitskii, S.: Generalized distances between rankings. In: Proceedings of the 19th International Conference on World Wide Web, WWW 2010, Raleigh, North Carolina, USA, 26–30 April 2010, pp. 571–580 (2010)

24. Kusner, M.J., Loftus, J., Russell, C., Silva, R.: Counterfactual fairness. In: Advances in Neural Information Processing Systems, pp. 4069–4079 (2017)
25. Lu, W., Miklau, G., Gupta, V.: Generating private synthetic databases for untrusted system evaluation. In: 2014 IEEE 30th International Conference on Data Engineering (ICDE), pp. 652–663. IEEE (2014)
26. Ma, S., Zheng, Y., Wolfson, O.: Real-time city-scale taxi ridesharing. IEEE Trans. Knowl. Data Eng. **27**, 1782–1795 (2015)
27. Markovsky, I.: Low Rank Approximation: Algorithms, Implementation, Applications. Springer, Heidelberg (2011). https://doi.org/10.1007/978-1-4471-2227-2
28. McFarland, D.A., McFarland, H.R.: Big data and the danger of being precisely inaccurate. Big Data Soc. **2**(2), 2053951715602495 (2015)
29. Meng, X., Li, H., Cui, J.: Different strategies for differentially private histogram publication. J. Commun. Inf. Netw. **2**(3), 68–77 (2017)
30. MetroLab Network. First, do no harm: Ethical guidelines for applying predictive tools within human services (2018, forthcoming). http://www.alleghenycountyanalytics.us/
31. Nabi, R., Shpitser, I.: Fair inference on outcomes. In: Proceedings of the AAAI Conference on Artificial Intelligence. AAAI Conference on Artificial Intelligence, vol. 2018, p. 1931. NIH Public Access (2018)
32. NYC Taxi and Limousine Commission. TLC trip record data (2018). http://www.nyc.gov/html/tlc/html/about/trip_record_data.shtml. Accessed 2 June 2018
33. Rastogi, V., Nath, S.: Differentially private aggregation of distributed time-series with transformation and encryption. In: Proceedings of the 2010 ACM SIGMOD International Conference on Management of data, pp. 735–746. ACM (2010)
34. Rubin, D.B.: Causal inference using potential outcomes: design, modeling, decisions. J. Am. Stat. Assoc. **100**(469), 322–331 (2005)
35. Salimi, B., Gehrke, J., Suciu, D.: Bias in OLAP queries: detection, explanation, and removal. In: Proceedings of the 2018 International Conference on Management of Data, pp. 1021–1035. ACM (2018)
36. Sweeney, L.: Discrimination in online Ad delivery. Commun. ACM **56**(5), 44–54 (2013)
37. Xiao, X., Wang, G., Gehrke, J.: Differential privacy via wavelet transforms. IEEE Trans. Knowl. Data Eng. **23**(8), 1200–1214 (2011)
38. Xiao, Y., Xiong, L., Fan, L., Goryczka, S.: Dpcube: differentially private histogram release through multidimensional partitioning. arXiv preprint arXiv:1202.5358 (2012)
39. Xu, J., Zhang, Z., Xiao, X., Yang, Y., Yu, G., Winslett, M.: Differentially private histogram publication. VLDB J. **22**(6), 797–822 (2013)
40. Zemel, R.S., Wu, Y., Swersky, K., Pitassi, T., Dwork, C.: Learning fair representations. In: ICML, pp. 325–333 (2013)
41. Zhang, Y., Thomas, T., Brussel, M., van Maarseveen, M.: Expanding bicycle-sharing systems: lessons learnt from an analysis of usage. PLoS One **11**(12), e0168604 (2016)
42. Zliobaite, I.: Measuring discrimination in algorithmic decision making. Data Min. Knowl. Discov. **31**(4), 1060–1089 (2017)

MODAL - A Platform for Mobility Analyses Using Open Datasets

Wender Zacarias Xavier$^{(\boxtimes)}$ and Humberto Torres Marques-Neto

PUC-Minas - Pontifical Catholic University of Minas Gerais,
Dom José Gaspar Street, 500 Coração Eucarístico, Belo Horizonte, Brazil
wender.xavier@sga.pucminas.br, humberto@pucminas.br

Abstract. Cities are becoming smart environments with the use of information and communication technologies (ICT). Data from these technologies are stored by various devices spread throughout the city and are available in open data portals, which can be used to improve essential services such as public transport and fed into platforms for visualization and analyses. Human and urban mobility analyses demonstrate that understanding movement patterns can assist governments in city's decision-making process, as well as improve life quality of citizens. Aiming to enable mobility analysis in different cities, this work presents *MODAL* platform. This platform replicates mobility analyses and algorithms on databases of different cities using data obtained from open data portals. We assess the platform with a case study performing analyses of the transportation displacement within three different cities using complex network metrics. The results demonstrated the public transportation system efficiency showing regions of Chicago, Dubai and Taichung well served and regions which are key points to the transportation city interconnecting various areas. Moreover, we could evaluate how improved the transportation system would be by adding new lines or new transport system. The analyses demonstrated the platform potential to be used as support decision system for governments, showing the possibility of applying open data to improve city services and facilitate the conduction of analyses on various cities.

Keywords: Human mobility · Urban mobility · Smart cities · Smart systems

1 Introduction

Cities are becoming increasingly intelligent with the use of information and communication technologies in areas such as sensing, security, health, transportation and human mobility [5]. These smart cities use such technology to improve urban infrastructure and life quality of citizens. With the advent of smart devices connected to the Internet together with more sophisticated computing, it is possible to develop an environment able to bring intelligent services [14]. Nowadays, these

J. Oliveira et al. (Eds.): BiDU 2018, CCIS 926, pp. 40–55, 2019.
https://doi.org/10.1007/978-3-030-11238-7_3

devices are used to create smart homes [6], smart grids [7], smart transportation [16,21], smart health systems [9], and culminating in smart cities [8].

Smart cities generates a considerable amount of data at high speed from different sources. This large volume of data is commonly known as Big Data and presents excellent potential as well as challenges for the development of smart cities [15]. In recent years emerged an increasing number of open data portals, making public a range of datasets. Many governments followed the trend mainly to make acts transparent to the population (e.g. United States Open Data Portal[1], Brazil Open Data Portal[2] and Africa Open Data Portal[3]). Some private companies have started following the same patterns providing part of the data for analyses and creation of new solutions by the public in formats of challenges and online tools (e.g., Uber[4], Yelp Challenge[5] and Telecom Italia Challenge[6]) [1].

Open data portals represent a valuable source of data both for citizens who can get information about public policies and services as well for governments who can use the data to improve city services and population well being [3]. Data available from these portals are applied in areas such as ecology [20], health [19], finance [4] transport, and human mobility [13]. The analyses conducted on human mobility patterns have potential in assisting governments and companies on the decision-making process in order to improve transportation efficiency [12].

Mobility and transport patterns are extensively studied to model and improve existing services [17,18]. From the Big Data generated by these applications, together with the advancement of computer processing power, several works began to interconnect databases of different types of transportation to obtain new insights and find patterns previously unknown [25,26]. The use of heterogeneous datasets assist on the investigation of new mobility inquires for the cities. Various mobility data are available through open data portals and are used to propose better solutions for transport services [23]. The replicability of these solutions can help understand and introduce better questionings for other cities assisting in the decision-making process by governments, improving life quality of citizens and reducing costs of transportation plans and implementation.

By using datasets available by open data portals, this works aims to present a platform capable of analyzing heterogeneous transportation data obtained by these portals, creating a supporting system for understanding human mobility on several cities. This platform called MODAL - Mobility Open Data Analyzer is capable of conducting mobility analysis with easy replicability on datasets of different towns, allowing big and small cities comprehend mobility within city boundaries. Moreover, it will enable the insertion of new functions and algorithms into the platform.

---

[1] https://www.data.gov/.
[2] http://dados.gov.br/.
[3] https://africaopendata.org/.
[4] https://movement.uber.com/.
[5] https://www.yelp.com/dataset/challenge.
[6] https://dandelion.eu/datamine/open-big-data/.

We performed analysis using complex networks metrics on open datasets of 3 distinct cities. These analyses demonstrated the potentials of the platform for dealing and conducting investigations on mobility using open data, and highlight that *MODAL* could be used to validate government decisions as well as replicate academia research results easily on various scenarios (e.g. land use, demographics and social studies).

This work is structured as follows. Section 2 presents some related work that motivates the development of *MODAL* platform. Section 3 presents the platform and explains the methodology used to conduct the analyses. Section 4 discusses the results of the proposed methodology and the platform potentials. The conclusions and future works are presented in Sect. 5.

2   Related Work

With the urban population growth, city services must be upgraded and rethought to improve life quality and make the city smarter. The smart cities market is expected to grow in the billions of dollars over the next few years by encouraging the development of new platforms and smart solutions for cities. This way, the work of Vilajosana et al. [24], approaches the main aspects to be observed in the development of platforms and solutions for companies and governments. The authors present a generic platform containing the main components for applications of intelligent cities using heterogeneous data. Finally, they show how the data flow generates useful knowledge for the corporate area, and how each actor in this flow generates and consumes the data.

Some papers focus on the use of data from a specific city to better understand its standards and to propose intelligent platforms. On D'Amico et al. [11], authors used data from geographic information systems (i.e., GIS) from the city of Naples in Italy to develop a decision support platform for sustainable city planning. To do this, they aggregated information from the city's public indicators to determine the regions where sustainable transport policies could best be applied to achieve a lower rate of pollution and emission of polluting gases. The work used as a basis the main challenges, objectives, and actions of the European mobility plan for the city of Naples and has proved to be a powerful tool to validate the criteria adopted by the government.

On the other hand, the work of Xu and González [25] integrates mobile phone and transportation databases to estimate the impact on traffic during the Olympic Games in Rio de Janeiro, in addition, to propose strategic solutions for a better dimensioning of public transportation for future events and the best strategy for collective mobility on an urban scale. The authors developed an iterative visualization tool to verify the time of travel among several areas of the city using various modes of transportation (e.g. bus, train and taxi).

The use of aggregate transport data assists in better planning and arrangement of the transport system in a city, but can also be used to understand how the use of the various modes of transportation can make mobility from one region to another more efficient. In Bahrehdar and Ghazi [2] the authors developed a decision support system for planning the urban journey considering various public transportation system. To do this, they used the bus, taxi, subway and BRT (i.e., Bus Rapid Transit) datasets, based on a proposed methodology applying graph algorithms to obtain a smaller path between two points. The authors determined the fastest route between city locations.

Finally, on Dur et al. [10] the authors developed a tool to provide valuable insight on the impacts on the development of a region. Details on land use, transport, urban services, population density, and pollution can be analyzed by the platform, which also has the visualization capabilities to help governments and other companies to become part of the concept of urban education in cities.

All these works presented above reveal mobility insights on the analyzed cities at different aspects. However, such analyses are focused and performed on specific cities with specific data, with no space to broad the analysis and visualization. *MODAL* tries to fill this gap being a platform where such analyses can be performed and validated in a practical way, focused on using open datasets due to its easy acquisition. Moreover, the platform enables the insertion of new datasets and algorithms by third parties, allowing the replication and new insights of human mobility.

3   Methodology

The methodology section is divided into two parts. First, we explain the development of MODAL and its architecture. Second, we present the methodology used to conduct the study case analyses.

3.1   MODAL

The Mobility Open Data Analyzer platform, *MODAL*, is a platform intended to replicate mobility analysis on different open datasets from various cities. Moreover, it has the potential to encourage publication and integration of new datasets for research and decision support purposes.

MODAL runs on a server at PUC-Minas and can be accessed through a web page. We developed using Python 3.5 programming language due to the broader use of data science projects with many libraries and mining algorithms. As the platform focus on conducting analyses on open datasets, all packages and libraries used to create the visualization are open source projects such as Open Street Map[7], Leaflet[8] and NetworkX[9]. At a given point the *MODAL* platform itself intend to be open source. However, these first phases of validation and standardization of analyses are done locally.

The platform is divided into three parts responsible for dealing with the data input, the analysis and algorithms, and finally, the visualization. Figure 1 shows a sample screen for the platform with three inserted datasets from the cities of

---

[7] https://www.openstreetmap.org/.
[8] https://leafletjs.com/.
[9] https://networkx.github.io/.

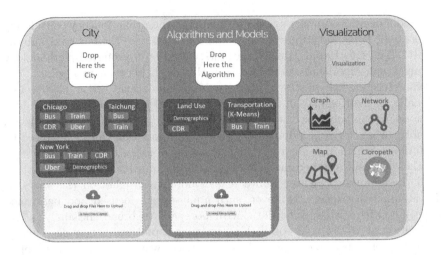

Fig. 1. MODAL platform overview.

Chicago, Taichung and New York and two available analyses which are land use and transportation analysis using complex network metrics.

The first module (i.e. section named City on the platform) comprehends the data input into the system. Once there are different datasets with different formats and structure on various open data portals of the cities, it is hard to deal with all kinds of data. In order to solve this issue, for each algorithm and model we will determine the standard structure of the input data, and at the moment when a user wants to upload data from a different source, the user can verify the structure of the uploaded file and parser to the correct structure for analysis. Different datasets can be uploaded to different cities and labeled according to the type of data (e.g., the city of Chicago contains bus, train, Call Detail Record (CDR) and Uber related data).

The Algorithms and Models are responsible for conducting the analyses on the input data. All the functions must be programmed in Python language, and the upload of new functions to the system will pass through a rigorous analysis of a commission to check the presence of malicious code and the value of the analysis. This way, the platform act as an open box to replicate and apply the analysis on different cities. Once the city data is selected, the user can choose the algorithm to perform analyses on the data. Each model shows the needed input data to work correctly (e.g., transportation analyses using K-Means algorithm requires bus and train related data).

Finally, the third part of the platform is responsible for creating the visualizations and showing results that can be used to support decision plans. The maps and results are displayed according to the selected algorithm applied and can be downloaded for further analyses.

3.2   Study Case Analysis

In order to perform an evaluation of the potentials and challenges of *MODAL*, we conducted analyses based on the work of [22]. This work was selected while performing a systematic literature review with keywords such as *smart cities* and *urban mobility*. The authors use clustering algorithms and complex network metrics to evaluate the transport system of the city of Curitiba in Brazil. With the clustering algorithm they detected regions that may be useful for public transport and the regions indicating potential congestion on the transportation system. The following modeling and metrics are based on this cited work.

The required information to conduct the analyses must contain information regarding public transportation, such as line name, and coordinates (i.e., latitude and longitude) for each station and bus stop. We used at this study comma separated value format as input file format.

The transport system of the cities was modeled as a complex network, composed of a graph $G = (V, E)$ where V and E are respectively the set of nodes and edges of the graph. Each station or bus point corresponds to a single node V while if a bus line or train line passes through two distinct nodes V, then an edge is created between these two nodes. All metro lines that are connected and the user can move in any direction without the need of leaving the station are considered one route for the purpose of this study. At the end, we get an undirected graph of the city's transport system. Two graphs were created for each city, the first (G_1) by V composed only of bus stops, and the second graph (G_2) with buses and other transportation systems. These two graphs were developed to verify the public transportation efficiency considering just one mode of transport (i.e., bus) and two or more (i.e., tram, metro).

To evaluate the connectivity and mobility between different regions of the city, and consequently between stations, we apply the K-Means clustering algorithm to determine city regions and groups of nearby stations. K-means clustering algorithm aims to partition N observations into K clusters in which each observation belongs to the cluster with the nearest mean, serving as a prototype of the cluster. After defining the k clusters, all nodes in each cluster i are replaced by a single node, which will be called *Supernode* in this work. All internal edges of each cluster were also removed. An edge connecting an external node j to a node belonging to a cluster i is replaced by an edge connecting j to the *Supernode* of the cluster i. This process is performed for all clusters of the complex network, resulting in a graph containing K *Supernodes*.

The following network metrics were calculated for each graph:

- **Average Degree:** Average degree of all nodes;
- **Path Length Distribution:** Histogram of all path lengths (given two *Supernodes* i and j, the path l between these nodes, $l_{i,j}$), corresponds to the smallest path to connect them;
- **Average Path Length:** Average of all path lengths;
- **Average Clustering Coefficient** (C)**:** Average of all clustering coefficients C_i (probability that a node i creates a complete graph with its neighbors);

– **Betweenness** (*Betwe*): Number of paths between all nodes to all other nodes passing through the analyzed node i;
– **Closeness** (*Closen*): Number of visited nodes from one source node to all other.

In terms of transportation system planning, degree based metrics indirectly defines how overloaded is a bus station or a region of the city. Path length metrics defines the distance of possible routes between two regions of the city, which directly affect the user's perception of the public transport efficiency. Clustering coefficient helps identify regions of concentration of bus stops and stations, which affect traffic around these and also give alternatives for user to connect to nearby points. Centrality coefficients (i.e. betweenness and closeness) aid in the identification of regions of great importance to the transport system, showing points that are on the smallest route to other regions of the city, as well as points that are close to all others given the availability of bus lines connecting two regions.

The metrics analysis can enhance the comprehension of public transportation systems within the city, evaluating the displacement and regions with concentration or lack of service. Such information are essential to improve city services towards a smart environment improving the users mobility and life quality.

4   Results

The analyses explained on the methodology aims to evaluate public transportation systems using complex networks metrics. To achieve this, it is required data containing the geographic location of public transportation stations and bus stops, with the corresponding identifier of the lines that serve each halt and station. Although some works rely on partnerships with companies to obtain such data, the required information is available on some open data portals.

We selected three portals which contained the required data. First, the Chicago Open Data Portal[10] provides information about all bus stops and rapid transit system (i.e., metro) lines and stations from 2013. Second, the Taiwan Open Data Portal[11] provides information regarding all bus lines and stops for the province of Taichung. The Taichung metro system was planned initially to contain five lines, however, due to political decisions the project was rejected, and the current plan contains three lines not finished yet. As the map of the initial plan is available online with the expected location of the stations, we created a dataset with the geographic position of these stations to verify how the Taichung transportation system could be improved with the construction of the metro system. Finally, Dubai Open Data Portal provides information about the bus, metro and tram systems of the city[12]. These three cities have a policy focused on policy transparency of decisions and public services, releasing data from various segments in their respective government portals.

---

[10] https://data.cityofchicago.org/.
[11] https://data.gov.tw/.
[12] https://www.dubaipulse.gov.ae/.

Table 1 shows an overview of the datasets. Chicago, Taichung, and Dubai are three large cities containing an extensive public transport system that covers all the boroughs or districts of the city. Despite the fact that the large Taichung area compared to Chicago, its population is smaller than the American city because much of Taichung's territory consists of mountains and vegetation. Dubai presents the bus system with the lowest number of bus stops despite the higher number of bus lines. Figure 2 shows the displacement of stops and stations in these cities. A few stops and stations are located outside the boundaries of Chicago and Dubai datasets.

After obtained the required information to conduct the analyses, we applied the K-means algorithms on the data. We used the machine learning library Scikit-learn[13] and set as input 100 iterations for each execution. Moreover, we defined K as the number of boroughs or districts of the city, that is, $K = \{77, 29, 13\}$ respectively for the cities of Chicago, Taichung, and Dubai. We set these values (i.e. k and number of iterations) empirically. Varying K alters the final result of the analysis, with greater K resulting in more subdivisions on the transportation system thus more detailed information, and lower K resulting on less subdivision and consequently overview of city regions. This variation can be easily done on the platform without the need of changing any code.

Once the clusters were defined, we proceed with the methodology creating the *Supernodes* and inserting the edges. Figure 3 shows the graph G_2 for the three cities. The edges thickness is related to the number of transport lines connecting the clusters. With this image, we see regions which are highly and poorly connected. On Taichung and Dubai images the more dense regions are downtown. On Chicago, due to a similar distribution of the public transportation lines, it is harder to determine the region with a higher concentration.

Table 1. Data overview for Chicago, Taichung and Dubai.

City	Chicago	Taichung	Dubai
Country	United States	Taiwan	United Arab Emirates
Area (km)	606.34	2,214.90	4,114
Population	2,792,164	2,695,598	3,052,000
Division	77 Boroughs	29 Districts	13 Districts
# Bus stops	11,011	7,049	4,456
# Bus lines	142	253	350
# Metro stations	141	95	49
# Metro lines	8	5	2
# Tram stations	–	–	11
# Tram lines	–	–	1

---

[13] http://scikit-learn.org/stable/modules/generated/sklearn.cluster.KMeans.html.

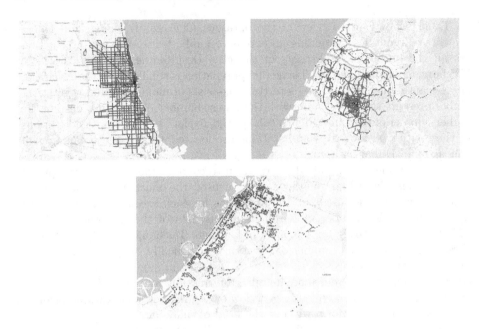

Fig. 2. Transportation systems distribution in the cities of Chicago (top-left), Taichung (top-right) and Dubai (bottom). In gray, bus stops, in red, train stations and in blue, tram stations. (Best seen in color)

With the graph created, we used the Python library NetworkX to calculate the metrics on our graphs. Analyzing Chicago data presented on Table 2, it is possible to see how the public transportation is improved with the use of the metro system. If the city had just bus routes, an average user living in a given region could reach approximately 20 other areas of the city, from a total of 77. Besides, to reach all other areas the user needed to take approximately two buses. Together with the metro lines, each region is reachable by approximately 35 regions of the city, 15 more compared to the bus. Moreover, it is needed an average of 1.6 transportation mode to reach all over regions. Adding these two transport system, it is possible to notice an increase on the clustering coefficient, showing that these two transportation systems improve the mobility on the city.

The same pattern can be seen on Taichung data, where regions are reachable by other 14.83 regions with the bus system, and approximately 3 more regions when the metro system would be added. These results demonstrate the importance of implementing the planned metro system on the Taichung city, moreover, could be used to verify and support city planners to identify important regions for new stops and stations.

Although adding two types of transport, tram, and metro, the Dubai system presented a small improvement of the overall mobility. By adding these two transport modes, we could see that just one edge was added to the graph, resulting in a slight increase in the city connectivity. As we are using the number of

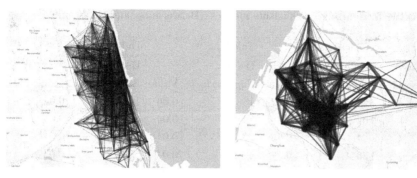

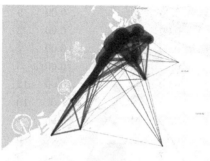

Fig. 3. The Graph G_2 for the cities of Chicago (top-left), Taichung (top-right) and Dubai (bottom). Line thickness is related to number of lines attending two nodes.

Table 2. Complex networks metrics computed for each city with bus (B), metro (M) and tram (T) datasets.

City	Chicago		Taichung		Dubai	
Dataset	B	B+M	B	B+M	Bus	B+T+M
#Nodes (*Supernodes*)	11,011 (77)	11,152 (77)	7,049 (29)	7,117 (29)	4,456 (13)	4,515 (13)
#Edges	755	1,330	215	277	63	64
Avg. degree	19.61	34.55	14.83	19.10	9.69	9.84
Avg. path length	1.94	1.60	1.47	1.32	1.19	1.17
Avg. clustering coef.	0.60	0.77	0.80	0.86	0.86	0.87

districts as clusters, using a higher number of clusters on Dubai city dataset could present new insights.

The importance of a region can be evaluated using metrics shown in Table 3, where we display regions who presented the five best results for closeness and betweenness, together with the five regions with the worst results. The number corresponding to the *Supernode* are presented on Figs. 4, 5 and 6.

Table 3. Centrality coefficients metrics. (Betweenness and clustering)

City	Chicago		Taichung		Dubai	
Dataset	B+M		B+M		B+M+T	
Betwe: 5+	*Supernode #*	Value	*Supernode #*	Value	*Supernode #*	Value
	1	0.42	18	0.08	10	0.04
	64	0.38	14	0.04	8	0.04
	4	0.35	27	0.04	0	0.04
	36	0.28	19	0.04	3	0.03
	26	0.24	6	0.02	4	0.02
Betwe: 5−	71	0.00	25	0.00	9	0.00
	9	0.00	20	0.00	2	0.00
	75	0.00	16	0.00	7	0.00
	52	0.00	26	0.00	5	0.00
	3	0.00	1	0.00	12	0.00
Closen: 5+	4	0.8	18	1.0	10	1.0
	65	0.79	14	0.96	8	1.0
	36	0.79	27	0.93	0	1.0
	12	0.77	19	0.93	4	0.92
	1	0.77	6	0.90	3	0.92
Closen: 5−	71	0.48	8	0.64	1	0.8
	52	0.47	1	0.62	7	0.8
	70	0.46	3	0.60	6	0.8
	34	0.45	17	0.58	12	0.7
	3	0.43	26	0.55	5	0.7

On Chicago dataset (see Fig. 4), clusters 1, 26, 64, 4 and 36 (located at south, west, northwest, downtown, and south-downtown respectively) presented higher values for betweenness, characterizing regions which bus stops belonged to most of the all shortest paths. Analyzing the region use, we could see that clusters 1 and 64 contains express highways widely used which connects different regions of the city including the downtown area. On the other hand, clusters 4, 36 (downtown regions), 12, 36 and 65 (south regions) are regions where bus stops were central to the whole public transportation system. These regions are well-served by the public transportation with strategical stops and stations. Opposite behavior is seen on clusters 3, 9 (located northwest) 52, 71 (located south) and 75 (west) which presented low values for betweenness. In addition, clusters 3, 52 and 71 showed low values for closeness metrics as well. These regions are located on the city boundaries of Chicago, characterized by lower efficiency of the public transportation.

Clusters near downtown presented high betweenness values for Taichung city (see Fig. 5). Clusters 14, 19 and 27 are relevant areas to reach other regions,

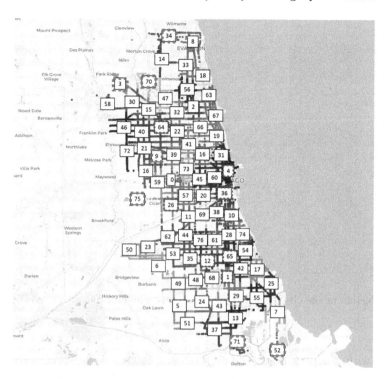

Fig. 4. Clusters for the city of Chicago with the corresponding *Supernode* id. (Best seen in color.)

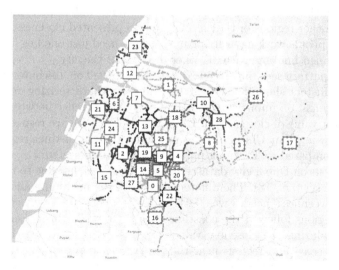

Fig. 5. Clusters for the city of Taichung with the corresponding *Supernode* id. (Best seen in color.)

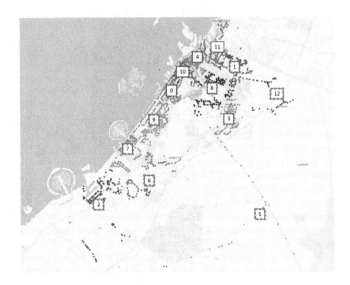

Fig. 6. Clusters in the city of Dubai with the corresponding *Supernode* id. (Best seen in color.)

together with the cluster 18 and 6. Taichung is crossed by two main rivers, Dajia and Daán river, and regions 18 and 6 are key clusters to connects crossed areas by these rivers. Due to this, these clusters present higher betweenness and closeness values. Moreover, cluster 6 is the region where the international Airport of Taichung is located, with many bus lines and stops connecting the airport with other regions of the city. Cluster 18 presented closeness value of 1, showing that people working or living in that area need just one bus line or train line (i.e., planned line) to reach all other districts of the city.

A similar pattern seen on Chicago datasets happened on Taichung city, where boundaries clusters showed low centrality metrics characterizing regions that need improvement on public services. However, the difference between the highest and lowest value of closeness for Chicago and Taichung are respectively 0.37 and 0.45, showing that Chicago transportation connects distinct regions somehow better compared to Taichung public transportation system.

The analysis on Dubai city dataset showed small variance due to the number of clusters we set (i.e., 18). The difference between the maximum and minimum value for betweenness metrics was 0.04. Dubai is a coastal city, where the division of the regions followed the coastline with some other clusters entering the territory. Analyzing the closeness values for this dataset, we see that regions on the south coast and regions near the north coast require more attention to improve transportation efficiency connecting all regions within the city. Clusters in this area (1, 7, 6, 12) requires a higher number of transport transfers to reach all regions of the city.

By using *MODAL* platform, we could perform the previous analysis on a practical way, varying the number of clusters and obtaining the visualizations and metrics values for the cities. In addition, we clearly see the valuable information we could obtain using open data, enhancing the comprehension of public services and evaluating government acts such as the plan of enlarge of a public transportation system.

5   Conclusions and Future Works

Cities are becoming smarter with the use of information and communication technologies, with a huge amount of data being generated by these devices and released on governments open data portal. These datasets present potential to improve city services, with information that could be used to develop applications aimed to support the decision-taking process of governments and enhance the life quality of citizens. This work presented the platform *MODAL* aimed to perform mobility analysis on a range of topics using Open Data.

The platform *MODAL* has the capabilities to replicate analyses on different datasets of various open data portals, while the results of these analyses can be used to improve urban mobility in big cities and improve life quality of citizens. In addition, the results proved the potentials of using data provided by open data portals not only to replicate analysis on other portals as well to validate results of analyses published on academic papers.

In order to check the potentials challenges and potentials of developing a platform supported by open data, we selected on the literature as a study case a work which used private transportation data to evaluate the public transportation of a city using complex networks metrics. At this point, with the goal set, we searched through open data portals to check the availability of the required data, as well as if the analyses could be conducted using the data. Although distinct open data portals often do not provide data on the same format or structure, the data containing information regarding coordinates of bus stops, train stations, and lines of service can be found on such portals. We selected three cities to conduct the analysis: Chicago, Taichung, and Dubai. These three cities have a policy focused on policy transparency of decisions and public services, releasing data from various segments in their respective government portals.

Applying a methodology based on the work of da Silva [22], we generated transport complex network and retrieved the results of some metrics applied to the graph. Then we verified city regions with better displacement of lines as well regions that could be improved to enhance the efficiency of public transportation and interconnect regions of the city. The replication of analyses on these three datasets and obtaining the results were conducted on a practical way on the *MODAL* platform. Moreover, we could aggregate data from transport map future plans, showing the power of the platform serving as a decision supporting system for governments reducing costs and evaluating plans prior to the construction.

As future works, we intend to conduct more analyses on *MODAL* and release the platform as open source. In addition, we intend to conduct scalability analysis using varied sizes of data. Moreover, we aim to provide more structured results so urban planners could add the information on governments official projects.

Acknowledgements. This work is supported by MASWeb (FAPEMIG/PRONEX APQ-01400-14), FAPEMIG (APQ-02924-16), PUC-Minas, CNPq, CAPES and STIC AmSud 18-STIC-07.

References

1. Attard, J., Orlandi, F., Scerri, S., Auer, S.: A systematic review of open government data initiatives. Gov. Inf. Q. **32**(4), 399–418 (2015)
2. Bahrehdar, S.A., Ghazi Moghaddam, H.R.: A decision support system for urban journey planning in multimodal public transit network. Adv. Railw. Eng. Int. J. **2**(1), 59–71 (2014)
3. Bakıcı, T., Almirall, E., Wareham, J.: A smart city initiative: the case of Barcelona. J. Knowl. Econ. **4**(2), 135–148 (2013)
4. Barkham, R., Bokhari, S., Saiz, A.: Urban big data: city management and real estate markets (2018)
5. Batty, M., et al.: Smart cities of the future. Eur. Phys. J. Spec. Top. **214**(1), 481–518 (2012)
6. Caragliu, A., Del Bo, C., Nijkamp, P.: Smart cities in Europe. J. Urban Technol. **18**(2), 65–82 (2011)
7. Chen, S., Song, S., Li, L., Shen, J.: Survey on smart grid technology. Power Syst. Technol. **8**, 1–7 (2009)
8. Chourabi, H., et al.: Understanding smart cities: an integrative framework. In: 2012 45th Hawaii International Conference on System Science (HICSS), pp. 2289–2297. IEEE (2012)
9. Demirkan, H.: A smart healthcare systems framework. IT Prof. **15**(5), 38–45 (2013)
10. Dur, F., Yigitcanlar, T., Bunker, J.M.: A decision support system for sustainable urban development: the integrated land use and transportation indexing model. In: Proceedings for the Second Infrastructure Theme Postgraduate Conference 2009-Rethinking Sustainable Development: Planning, Infrastrucutre Engineering, Design and Managing Urban Infrastructure, pp. 165–177. Queensland University of Technology (2009)
11. D'Amico, P., Di Martino, F., Sessa, S.: A GIS as a decision support system for planning sustainable mobility in a case-study. In: Ventre, A., Maturo, A., Hošková-Mayerová, Š., Kacprzyk, J. (eds.) Multicriteria and Multiagent Decision Making with Applications to Economics and Social Sciences, pp. 115–128. Springer, Heidelberg (2013). https://doi.org/10.1007/978-3-642-35635-3_10
12. Ferraris, A., Santoro, G., Papa, A.: The cities of the future: hybrid alliances for open innovation projects. Futures **103**, 51–60 (2018)
13. Ferster, C.J., Fischer, J., Manaugh, K., Nelson, T., Winters, M.: Using OpenStreetMap to inventory bicycle infrastructure: a comparison with open data from cities. Technical report (2018)
14. Hashem, I.A.T., et al.: The role of big data in smart city. Int. J. Inf. Manag. **36**(5), 748–758 (2016)
15. John Walker, S.: Big data: a revolution that will transform how we live, work, and think. Int. J. Advert. **33**(1), 181–183 (2014)
16. Kim, H.-J., Lee, J., Park, G.-L., Kang, M.-J., Kang, M.: An efficient scheduling scheme on charging stations for smart transportation. In: Kim, T., Stoica, A., Chang, R.-S. (eds.) SUComS 2010. CCIS, vol. 78, pp. 274–278. Springer, Heidelberg (2010). https://doi.org/10.1007/978-3-642-16444-6_35

17. Mandl, C.E.: Evaluation and optimization of urban public transportation networks. Eur. J. Oper. Res. **5**(6), 396–404 (1980)
18. Manser, P., Becker, H., Hörl, S., Axhausen, K.W.: Evolutionary modeling of large-scale public transport networks. In: 97th Annual Meeting of the Transportation Research Board (TRB 2018). The National Academies of Sciences, Engineering, and Medicine (2018)
19. McBride, K., Aavik, G., Kalvet, T., Krimmer, R.: Co-creating an open government data driven public service: the case of Chicago's food inspection forecasting model. In: Proceedings of the 51st Hawaii International Conference on System Sciences (2018)
20. Reichman, O.J., Jones, M.B., Schildhauer, M.P.: Challenges and opportunities of open data in ecology. Science **331**(6018), 703–705 (2011)
21. Sherly, J., Somasundareswari, D.: Internet of Things based smart transportation systems. Int. Res. J. Eng. Technol. **2**(7), 1207–1210 (2015)
22. da Silva, E.L.C., de Oliveira Rosa, M., Fonseca, K.V.O., Luders, R., Kozievitch, N.P.: Combining k-means method and complex network analysis to evaluate city mobility. In: 2016 IEEE 19th International Conference on Intelligent Transportation Systems (ITSC), pp. 1666–1671. IEEE (2016)
23. Stone, M., Aravopoulou, E.: Improving journeys by opening data: the case of transport for London (TfL). Bottom Line **31**, 2–15 (2018)
24. Vilajosana, I., Llosa, J., Martinez, B., Domingo-Prieto, M., Angles, A., Vilajosana, X.: Bootstrapping smart cities through a self-sustainable model based on big data flows. IEEE Commun. Mag. **51**(6), 128–134 (2013)
25. Xu, Y., González, M.C.: Collective benefits in traffic during mega events via the use of information technologies. J. R. Soc. Interface **14**(129), 20161041 (2017)
26. Xu, Y., Li, R., Jiang, S., Zhang, J., González, M.C.: Clearer skies in Beijing-revealing the impacts of traffic on the modeling of air quality. Technical report (2017)

Urban Sensing

MENSAGERIA: A Smart City Framework for Real-Time Analysis of Traffic Data Streams

Marcos Roriz Junior[1,2], Rafael Pereira de Oliveira[1], Felipe Carvalho[1],
Sergio Lifschitz[1(✉)], and Markus Endler[1]

[1] Departamento de Informática,
Pontifícia Universidade Católica do Rio de Janeiro (PUC-Rio),
Rio de Janeiro, RJ, Brazil
{mroriz,rpoliveira,fcarvalho,sergio,endler}@inf.puc-rio.br
[2] Engenharia de Transportes – Faculdade de Ciências e Tecnologia,
Universidade Federal de Goiás (UFG), Aparecida de Goiânia, GO, Brazil
marcosroriz@ufg.br

Abstract. Several smart city systems have focused on addressing a specific mobility problem scenario (*e.g.*, air pollution, traffic jam) in a given city. The task of adding, extending, or porting the smart city scenario to other cities can be very challenging due to the rigid structure of such existing systems. To address this issue, in this paper we investigate common programming constructors that can be used to leverage the construction of such dynamic, smart city systems in the mobility domain. We propose MENSAGERIA, a framework based on both the Complex Event Processing data-streaming processing paradigm and relational database management systems, which can dynamically deploy new or extend existing smart city scenarios in near real-time and maintain an updated dataset for provenance purposes. MENSAGERIA provides several real-time primitives, such as *filter*, *join*, and *enrich*, that can be used to integrate, process, and analyze the city entities data streams. We discuss the generality, performance, and limitations of the proposed constructs through a real-world case study that was used in the Olympic Games of Rio in 2016 to detect, in real-time, existing and new situations that could affect the city mobility infrastructure.

Keywords: Smart city · Data stream processing · Urban computing

1 Introduction

The idea of Smart Cities has recently gained importance due to the growing need to address urban problems and use city resources more efficiently, mainly by collecting and analyzing data, automating some processes and employing intelligent decision making.

To cope with the ever-changing nature of Smart Cities scenarios, city operators need robust systems to handle each problem situation rapidly. For example,

© Springer Nature Switzerland AG 2019
J. Oliveira et al. (Eds.): BiDU 2018, CCIS 926, pp. 59–73, 2019.
https://doi.org/10.1007/978-3-030-11238-7_4

in preparation for the Olympic Games of 2016, the city of Rio de Janeiro required a smart city system to detect mobility risk scenarios. These risks were defined *a priori* (*e.g.*, traffic jam near an Olympic arena) and those not foretold (*e.g.*, an accident in a distant metro station that impacts the entire city infrastructure). To keep up with the city dynamicity, these smart city systems needed to be able to describe and detect new situations on the fly. It can be useful to detect not only foretold situations but also to further analyze and inspect the causes of an existing problem.

Moreover, such smart city systems need to be able to integrate, and timely process data streams originated from multiple sources within a few seconds. For instance, to timely respond to event situations, the Olympic Games smart city system need to correlate in real-time messages coming from different transport modalities and social network data streams, such as the city bus, metro, and railway fleets and the Waze and Twitter social network. Furthermore, in parallel, such systems need to store and maintain all the data stream rightly in a database to deploy analytics queries and extract strategic information.

Motivated by the limitations of current smart city systems concerning real-time dynamicity and generalization, this research paper aims to investigate common programming constructors that can be used to leverage the construction of such dynamic, smart city systems. To do so, we propose MENSAGERIA, a framework based on the Complex Event Processing (CEP) data-streaming processing paradigm [16] that can process and deploy existing and new smart city problem scenarios in real-time. CEP provides a set of online and real-time primitives through continuous query, such as *filter, join, enrich* and *negation*, that can be used to process and analyze data streams. Our main idea is to extend and expose CEP primitives as higher-level constructors to enable developers to define and alter several smart city scenario situations on the fly.

We validated the MENSAGERIA *framework* constructs by using it in a real-world case study during the entire Olympic Games of Rio in 2016. There, MENSAGERIA was used to detect and monitor in real-time mobility problems that could affect the Olympic Games. To do so, MENSAGERIA merged and analyzed the data streams messages coming from multiple mobility entities (*e.g.*, bus, metro, waze) in the Smart City. On the other hand, the city's engineers could describe the mobility problem scenarios' using the framework constructs. Concerning performance, MENSAGERIA was able to detect the mobility problem scenarios in real-time and engineers were able to deploy new problem situations within a few seconds.

The remainder of the paper is structured as follows. Section 2 motivates the research problem and overviews the fundamental concepts used throughout the paper. Section 3 discusses the MENSAGERIA framework architecture components and its implementation. Section 4 discusses a case study of the middleware usage in the Olympic Games of Rio in 2016. Section 5 compares our approach with other smart city frameworks and middleware platforms, while Sect. 6 summarize the concluding remarks and our plans for future work.

2   Context and Fundamental Concepts

Regarding software systems, many authors define a Smart City as the integration of social, physical, and IT infrastructure to improve the quality of city services. Despite the existence of several smart city initiatives in different countries, these deployments are often based on custom systems that are neither interoperable nor portable across cities or cost-effective [7,14].

To build such dynamic, smart city systems, the research community must address critical challenges in the areas of Networking, Database, High-Performance Distributed Computing, Software Engineering, Data Analysis, among others. Among such problems is how to describe and deploy new detection situations on the fly while receiving and process the data stream, that is, without interfering with the current system operation. In addition, given the massive data stream volume generated by the city's entities, one of the main issues is how to monitor and process the large incoming data stream generated by the city entities within few seconds. Timely processing such large data streams can be troublesome, especially considering high data stream velocities (throughput).

In this section, we will focus on two of the four Smart City elements: Big Data and Data Stream Processing, where for the latter we will describe and justify the use of a specific stream processing called Complex Event Processing (CEP) [15]. CEP is a compelling paradigm that is suitable for checking - in run-time - the occurrence of patterns among widely different events, patterns that may specify temporal, causal, spatial or semantic relation.

2.1   Big Data in Smart Cities

Most authors consider Big Data as a set of techniques and tools to store, process and manipulate large and incomplete data sets whereas conventional storage technologies, such as relational databases and sequential processing tools cannot deal with such a vast volume of data. According to [6,8] there are four major 'V' characteristics of Big Data: Large **V**olume; High data **V**ariety; High Processing **V**elocity and need of Data **V**eracity.

Big Data tools are already used by Smart City platforms, including NoSQL databases such as MongoDB and HBase; parallel data processing tools [19], such as Apache Hadoop and Apache Spark; real-time data stream processing tools [12], such as Apache Storm, Spark and Complex Event Processing; and visualization tools [1], such as Rapid-Miner.

Al Nuaimi *et al.* [2] discuss potential applications of Big Data tools in Smart Cities, such as recognizing traffic patterns and using historical data to locate the causes and avoid traffic jams. It facilitates the decision making of city governments using analyses of large data sets, and predicting the use of resources, such as electricity, water, and gas, in different situations using historical and real-time data.

Many authors, such as [1], have advocated combining IoT and Cloud Computing, coining the term "Cloud of Things." Their idea is to store and process all the data from an IoT network in a cloud computing environment, which is currently used in some Smart City projects [13,18,22].

2.2   Complex Event Processing

Complex Event Processing (CEP) is a programming paradigm that supports reactions to a stream of event data in real time [11,15]. In contrast to database management systems, in which data is first stored and then queried later, CEP stores continuous queries and runs data through them, *i.e.*, rather than storing the data, CEP focuses on continuously analyzing and processing the data whiles it passes, using the saved queries. Each continuously query implement one or more CEP real-time primitives, such as *filter*, and *negation*.

An event data stream is the resulting sequence of events created and sent by producers [10,15]. A continuous query uses its real-time primitives, such as *filter*, *split*, *project*, *sequence*, and *negation*, to react and to process the incoming event data stream as it passes. For instance, filter the *LocationUpdate* event stream to discover moving objects that are close to a given point of interest.

Continuous queries output events are known as a complex event since they represent higher-level information [15]. As an example, a complex *TrafficJam* event can be built by combining multiple *LocationUpdate* events in the same area and period. It is possible to create hierarchies of events, in which intermediate events can be used to further define other higher-level complex events.

There are several CEP engines implementation, for example, Esper [9] and Apache Flink [5]. The main difference between such engines are the language constructs used to define the event types and the supported real-time primitives. Most of CEP engines are based on the Continuous Query Language (CQL) [3] stream-oriented languages due to its formalism and its similarity with the SQL language.

3   Architecture

Smart cities systems seek to provide an accurate digital model of the real world to enable better planning and improve real-time decision making in response to problems. To support this decision making, such systems can make use of the city's historical data streams to design and test new CEP continuous queries in simulated scenarios before applying queries to the streams generated in real time. It can be useful to analyze beforehand the impact of strategic and operational decisions.

Building systems that can simultaneously handle such dual (historical and real-time) processing is a changeling task. A possible solution is to develop separate systems for each processing modus. However, as explained in [7,14], the process of replicating, adapting and integrating multiple smart systems is complex, costly and time-consuming.

Addressing these specific requirements, the MENSAGERIA architecture is designed to provide uniform programming constructs for handling data stream and persistent static data. They allow users to describe and correlate data streams and static data interchangeably.

To support these two types of data processing, the MENSAGERIA framework provides programming constructs that combine CEP primitives for real-time stream processing, with RDBMS functions for handling off-line data. The primary rationale of using CEP to provide MENSAGERIA's real-time constructs is due to its vast range of processing primitives and the flexibility of most CEP engines to deploy event pattern rules on the fly. Further, CEP provides many types of windows, e.g., Landmark, Sliding and Fading windows [3,17], each of which defines specific ways to select events for analysis of the virtually unbound data stream.

Complementary, MENSAGERIA also provides offline constructs to store, transform, and correlate streaming data with/to static data (e.g., points of interests, historical data, etc.). The main reason for such decision is the ability to store large data sets, e.g., relevant parts of the data stream for further analysis, with consistency guaranteed through ACID properties, even in the event of system failures. Hence, MENSAGERIA joins two well-established technologies, namely CEP and RDBMS, as a strategy to perform both real-time and non-real-time tasks.

3.1 API Integration and Database Schema

To provide uniform constructs, MENSAGERIA generalizes the concept of received and stored events into *messages*. The schema illustrated in Fig. 1 describes the structure of this and correlated entities in the role of data integration.

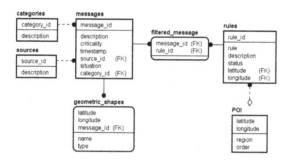

Fig. 1. Schema of the MENSAGERIA entities.

Each incoming event *should* contain the following attributes. First, it needs to include a source_id to identify the event producer. Knowing the source is useful because an event may trigger a situation that requires an action, such as remote actuation or dispatching some team for help. Similarly, events should include the position through the latitude and longitude fields to provide additional information as to where it occurred. This is especially useful for moving event producers, such as vehicles and city agents. Besides, a timestamp field is required to represent when it happened because a producer may postpone the transmission of an event to reduce the battery usage.

The events also require a `category_id` field, where it identifies the data type of the incoming event (*e.g.*, traffic jam notifications, road accidents, road maintenance). System users, such as business management and city operators, define these categories. Further, events also include a `title` and `description` field that contains respectively the summary and data payload. For instance, an event indicating a traffic accident, can include the text "`Accident on road 101`" as `title`, and "`Accident on 101. No fatal victims. Two Injured`" as `description`.

Similarly, events should include status information, through the `situation` and `criticality` fields. Situation indicates whether the occurrence of such an event is still valid, *e.g.* if a traffic road blockage is still happening or have been already solved. On the other hand, criticality is a scale of how severe is the given event to the proper functioning of the smart city system. For example, for a given smart city domain, the criticality of an event describing that a bulb in a street light is not correctly working is lower than a traffic accident that can injure citizens and influence the overall city mobility. In addition to such fields, MENSAGERIA will automatically attribute a globally unique `message_id` to the incoming event.

As can be noticed in the architecture diagram (Fig. 2), the MENSAGERIA framework has multiples API readers to collect message from external sources. These sources can use different data structures, entities or field names. Hence, we use the provided database schema definition to provide a unified view of them. Such data integration is necessary to decrease the complexity of writing CEP EPAs, as it is easier to write continuous queries (rules) over a well-defined and fixed data structure. This normalization works as a layer between the external stream sources and the CEP rule engine, dealing with data structures variations and the inclusion of new external sources. For example, in our experimental Olympic Games case (see Sect. 4), we extended these APIs several times to increase new data sources into the message stream, and not all of them resulted in modifications on processing and storing framework layer.

Beyond the data stream, the database schema was designed to store the rules and the Point Of Interest (POI) referenced by a subset of these rules. These POIs can be used to define those critical points to monitor in some aspect. For example, it is possible to check transport hubs like subway stations or bus stations concerning security issues or over crowed through the use of references to POIs by rules. If an incoming message contains a text containing "accident" and is within the point of interest (POI) location a complex `AccidentInSubwayEvent` is created, as described in Code 1.1.

```
1  INSERT INTO AccidentInSubwayEvent
2  SELECT msg.description, msg.latitude, msg.longitude, "4" as SEVERITY
3  FROM MessageStream AS msg, POI AS poi
4  WHERE msg.contains("accident")
5    AND distance(msg.latitude, msg.longitude,
6                 poi.latitude, poi.longitude) <= 100
7    AND poi.description like 'subway station'
```

Code 1.1. Example of rule using POI

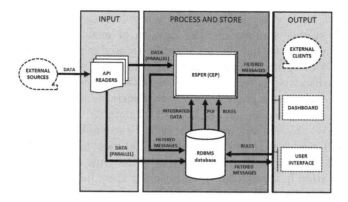

Fig. 2. Architecture diagram of MENSAGERIA

MENSAGERIA allows rules to use stored POIs to process messages related to one or multiple sites. This type of filtering can be done by direct reference the desired latitude and longitude into the rule, but we considered that and discarded it. Firstly, we abandoned it because managing the latitude and longitude directly on the rules seems to be laborious. Secondly, because it could increase the maintenance work significantly – imagine multiples rules filtering messages related to itinerant events. And lastly, because it was possible to group multiple points using stored POIs and apply the same rule to all of them. For example, in Code 1.1 it is possible to run the same rule to all subway stations previously registered with the description *'subway station'*. So, when there are new stations or some are disabled, this rule does not need to be updated.

Another feature provided by this database is that it stores the provenance about which rule filtered a determined message or trigger some action. The relation 'filtered_message' saves all match between rules and messages. It can be used to posterior analyses over the streaming processing progression and audit purposes.

3.2 Data Stream Processing

One of the basic constructs of the MENSAGERIA framework are API readers, illustrated in Fig. 2 in the input layer. Developers use this construct to instruct the framework on where and how to retrieve the input data stream. Each API reader implements the specific logic to consume the corresponding stream. Precisely, this is done by implementing a interface method named getMessages(). In addition, each API reader needs to normalize the incoming data to the message schema, that is, the output of the getMessages() method should be a collection of messages. After creating the API readers, developers register them into the framework, which is essential because the MENSAGERIA framework periodically activates these API readers to retrieve the input stream.

After the messages are retrieved, the framework starts the processing and storage cycle. We save each incoming message into the database table message.

Our framework enables developers to specify how long the data should be stored to cope with storage limitations. The lifespan of messages is specified when creating the database. By default, we discard no stored messages in the database.

Parallel to saving incoming messages, the framework executes the processing cycle. First, it discards any remainder message that is in the CEP engine memory. The primary decision for doing so is to avoid message duplication. Such issue appears when an API reader pulls a chunk of the data stream from a given source that contains a subset of the previously pull. To exemplify this issue, consider a Twitter API reader that retrieves the last 100 messages of a given hashtag. Consecutive API reads may contain a subset of the same messages, which would lead to duplication in the stream when injecting them into the CEP engine.

After cleaning the CEP memory, the framework injects incoming messages into the engine, which in turn correlate them with its existing rules (EPAs). If there is a match, both the filtered message and the corresponding EPA rule are sent to the output layer. As said before, and exemplified by Code 1.1, rules describe one or more condition that will trigger a given action. These EPAs can be chained to monitor complex patterns, for instance, a collection of events indicating that a vehicle is slow can be used to trigger a traffic jam event. In these situations, all filtered messages and triggered EPAs in the chain will be sent to the output layer.

Periodically, the framework retrieves all rules that are stored in the database to synchronize them with the CEP engine. If the retrieved rules are not present in the CEP engine, the framework instantiate the EPA rule. Similarly, if a rule is no longer present in the database and is present on the engine, the framework destroys the corresponding rule EPA.

After the rules update, the framework injects the incoming message to the CEP engine. The engine process and tries to correlate the incoming message with the existing rules. If there is a match, the filtered message and the corresponding EPA rule are sent to the output layer. The specified rule correlates them. Filtered messages are directed to the output layer, while the remaining ones stay in the memory for a brief time (as determined by developers).

Parallel to saving incoming messages, the framework executes the processing cycle. We may split this cycle into two phases: update and execution. Periodically, the framework retrieves all rules that are stored in the database to synchronize them with the CEP engine. If the retrieved rules are not present in the CEP engine, the framework instantiate the EPA rule. Similarly, if a rule is no longer present in the database and is present on the engine, the framework destroys the corresponding EPA rule.

3.3   Implementation

MENSAGERIA was implemented using the Java programming language and several open source libraries. The reason for using Java is due to the numerous open source libraries, framework, and middleware platforms available in this language. For example, we opted to use the Esper CEP Engine [9], one of the leading open source CEP engines, which is available as a Java library. Esper provides a continuous query CQL-like declarative language that supports CEP's transformation and pattern-based primitives. Further, Esper provides the ability to deploy new EPA rules dynamically. We extend this feature to support the ability to create and remove new rules dynamically. On the other end, we chose the RDBMS PostgreSQL to store and query historical messages and rules. The reason for this choice is due to PostgreSQL being an open source project and known for its large-scale data distribution capabilities. For the implementation of our proposal, we chose the RDBMS PostgreSQL version 9.5. Mainly, because it is an open source project, free, and meet all our requirements regardless of data distribution capabilities.

4   Case Study

To exemplify MENSAGERIA's programming concepts, this section describes a real-world case study involving its usage by the Rio Operation Center (*Centro de Operações do Rio* – COR) in the Olympic Games of Rio de Janeiro in 2016, as illustrated in Fig. 3. The framework was instantiated to interconnect, manage and process heterogeneous data streams from multiple sources, such as those originated by the city's bus fleet, the city's police force, the city's stations, the Olympic agents, transit applications (*e.g.*, Waze[1], Moovit[2]) and social networks (*e.g.*, Twitter[3]). By using the framework, COR aimed to detect in real-time (*few seconds*) changes and threats to the city mobility and security, especially those that could directly affect the Olympic Games.

To accomplish this task, COR had to integrate the multiple data stream produced by the city's elements, such as bus fleet, subway stations, and city agents. Further, this integration needs to handle external data sources, such as transportation applications and social networks.

To correlate such messages, there was a need to normalize the incoming data streams as messages. By doing so, it enables the city's operator to create processing rules that can detect specific scenarios. To cope with unexpected scenarios, the operator needed the ability to deploy new rules in real-time. Further, to prepare or learn different city scenarios, COR required the ability to simulate new rules with historical data that, in turn, requires that the framework to store and be able to replay data streams.

As shown, the Olympic Games case offered a challenged smart city scenario to MENSAGERIA. For example, to monitor a car crash or traffic jam in a critical street it needs to integrate the messages from bus fleet, transit applications like Waze and Moovit, internal communication channel agents and messages with specific *hashtags* from social networks. To do so, COR used several filters to correlate and process such messages. For instance, it used latitude and longitude points to specify polygons and define points of interest (POI) during the Games.

---

[1] Waze – https://www.waze.com/.
[2] Moovit – https://moovit.com/.
[3] Twitter – https://twitter.com.

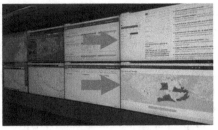

(a) COR's management center   (b) COR's MENSAGERIA instances

Fig. 3. MENSAGERIA usage in Centro de Operações do Rio (COR)

So, MENSAGERIA uses the messages' geographic coordinates as criteria during the query execution and the priority of each message in the data stream.

4.1   Performance

One of our contributions is to propose a mixed architecture to process messages in real time while correlating the messages with static data and further storing them in a structured and recoverable way. The purpose of storing parts of the event stream is not only to correlate with incoming messages but be able to reconstruct past processing window for learning and validating intends.

To reconstruct such timeline efficiently, we decided to use an RDBMS instead of non-relational databases, such as flat data files. We were aware that it could be quicker save the stream without structure, validation or constraints. Besides, our experimental results show that (see Fig. 4). However, we tried to construct a tool to support real-time decision making in smart city scenarios where the framework's processing constructs aim for robust and reliable data storage systems. For instance, the use of RDBMS provided the MENSAGERIA to (i) use standard analytic RDBMS tools to explore the stream in real time; (ii) facilitate correction of historical data on event stream; (iii) integration of pre-existing tools for processing and exploring stored stream data.

The experiments focused on define establish performance boundaries for our prototype. We performed tests in four scenarios: (i) only processing the data stream without saving it ("No Saving" in Fig. 4); (ii) processing the data stream and saving it in NoSQL flat data files; (iii) saving it using our proposed architecture ("RDBMS Parallel Saving" in Fig. 4); and (iv) processing the data stream and saving it sequentially and immediately ("RDBMS Sequential Saving" in Fig. 4).

For each scenario, we measured the framework's throughput, concerning the time to process and save inputs with a size between 100,000 and 2,500,000 messages. To approximate our tested scenarios to real data stream conditions, we used real messages collected from Twitter [24].

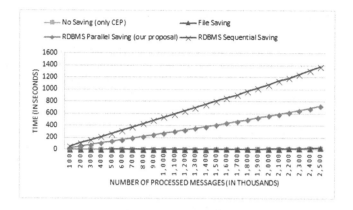

Fig. 4. Performance test of our architecture MENSAGERIA

The machine used during the experiments was an off-the-shelf computer with Intel Core I5 (2.3 GHz), 8 GB RAM, and 512 GB Hard Disk. We have implemented with Java, and we used the PostgreSQL 9.3 RDBMS.

The results on Fig. 4 shows evidence that our parallel approach performs better than the sequential method, and support our architectural choice of parallelly feed the RDBMS and the CEP engine ESPER [9] (see Fig. 2).

Another finding was that the Framework could process a reasonable volume of messages using an ordinary machine. We were able to integrate, process, filter and store 2.5 millions of messages in 12.2 min from different external sources. It makes an average of 3,416 messages per second using this setup to run both the MENSAGERIA and the RDBMS. This absolute value may not deal with massive social network data. However, if we consider other hardware and database tuning possibilities, our approach is encouraging. Nevertheless, it was tested to solve a real-world problem during the 2016 Olimpic Games performing agreeably and fulfilling all expectations, showing capable of handling large volumes as well when using more robust hardware infrastructure.

5   Related Works

The challenge of dealing with the real-time processing of high volumes of data in Smart Cities ecosystems is widely discussed and investigated in research. In this section, we argue five topics for comparison of solutions involving real-time data analytics in a Smart City scenario. These topics, listed in Table 1, are: (i) capacity for dealing with online (real-time processing) and offline (batch processing) analytics; (ii) the stream processing technology used in the solution and whether they provide storage capabilities; (iii) if the rule deployment mechanism is dynamic, able to deploy at run-time, or static; (iv) the architecture of the solution, i.e., distributed or centralized; (v) the data processing order, indicating whether there is a sequential or parallel data storage procedure, regarding the stream processing.

The authors of [23] present a Smart City framework, namely CityPulse. This framework provides real-time stream processing and large-scale data analytics, although they do not mention the technology used for that. It also provides data enrichment with semantics annotations, which enables adaptive processing, aggregation, and federation of data. The paper does not emphasize whether the framework can store data in an RDBMS or NoSQL data store.

In [4], the ALMANAC framework is presented. The authors explore the use of a block-based approach to composing real-time data processing rules for Smart Cities. The primitives for rules are written in JSON syntax and composed in blocks, which encapsulate them. These blocks represent monitoring and alerting tasks as reusable and modular "processing chains." They use a CEP engine for the stream processing and rules can be deployed dynamically.

The authors in [12] propose a Smart City framework for dealing with Big Data issues of real-time stream processing. The prototype is built using Apache Storm, which offers CEP and storage capacities for dealing with data. The architecture uses a distributed OpenNebula Cloud infrastructure which provides scalability. When we inject data into the framework, it is enriched with contextual information using an ontology-driven approach.

In [6] the authors present the CiDAP (City Data and Analytics Platform) which was tested in the SmartSantander [20] testbed. CiDAP is capable of doing real-time and also offline processing of data. For that, they use a CouchDB database that receives all data that is later processed using Spark, a CEP solution developed by Apache. It is a distributed architecture and rules can be deployed at run-time.

Table 1 presents a comparison between related works and MENSAGERIA system. As shown in Table 1, MENSAGERIA can do real-time and offline analytics with the proposed architecture. It also allows the deployment of rules at run-time. The main point of comparison is that data is stored in an RDBMS and processed in parallel through the CEP (Esper) engine.

Table 1. Comparison between related works and MENSAGERIA

	Analytics	Stream processing technology	Rule deployment	Architecture	Data processing
CityPulse	Real-time	Not mentioned	Dynamic	Not mentioned	Sequential
ALMANAC	Real-time	CEP + Blocks	Dynamic	Distributed	Sequential
Girtelschmid et al. [12]	Real-time	Apache Storm	Dynamic	Distributed	Sequential
Cheng et al. [6]	Real-time/offline	CouchDB + Apache Spark	Dynamic	Distributed	Sequential
MENSAGERIA	Real-time/offline	RDBMS + CEP (Esper)	Dynamic	Distributed	Parallel

6   Conclusions

Several cities have started to introduce advanced ICT-based services, including Internet of Things, Big Data analytics and Stream Processing for better planning and improvement of the situational awareness of and reactivity to city problems such as in traffic jams, public transport bottlenecks, localized missing public services (e.g., garbage collection) and emergency response. Such planning and awareness are particularly necessary when the city hosts a mass-event, such as Olympic Games, a religious or cultural festival (e.g., RockInRio), New Year's Eve, etc., where many city points become hotspots and traffic hubs, and where the public transportation system receives an additional and abnormal number of passengers.

This paper describes the MENSAGERIA framework which combines Complex Event Processing (CEP) primitives for real-time stream processing with RDBMS functions for the off-line handling of the sensor and social data. The MENSAGERIA was custom-tailored for the Centro de Operações do Rio (COR) that required a tool for straightforward definition and on-the-fly deployment of processing rules for data streams as well as off-line analysis of data stored in an RDBMS.

Hence, the main contributions of this paper are:

1. Design and implementation of the MENSAGERIA framework with simple-to-use constructs that can be tailored to detect smart city situations in real-time, and also deploy new situations on the fly, from the data streams produced by many types of city entities;
2. An ingenious approach to combine data stream processing and stream persistence and formulate queries and rules uniformly;
3. Validation and evaluation of the proposed framework in a real-world case study involving the Olympic Games of Rio in 2016, and its continued use by COR after the games.

Since it went "into the wild" in early 2016, the MENSAGERIA was well incorporated into the work routine of COR personnel and was fundamentally helpful for detecting (i) vehicle accidents in sensitive places like Rebouças Tunnel[4], Linha Vermelha[5]; (ii) train malfunctions or overcrowded before, during and after the games; and (iii) any transit anomaly inside predetermined areas (usually surrounding all Olympic Game Sites).

We may still improve MENSAGERIA in the following ways: (i) The data storage cost presented during the experiments can decrease using database tuning techniques like the creation of access structures such as indexes or materialized views [21]. Another option is to distribute the dataset throughout multiple machines using the RDBMS interfaces without any impact on our framework.

Acknowledgments. The authors would like to thank Alexandre Cardeman and Dario Bizzo Marques from Centro de Operações do Rio de Janeiro (COR).

---

[4] http://pt.wikipedia.org/wiki/Túnel_Rebouças.

[5] https://pt.wikipedia.org/wiki/Linha_Vermelha_(Rio_de_Janeiro).

References

1. Aazam, M., Khan, I., Alsaffar, A.A., Huh, E.N.: Cloud of things: integrating Internet of Things and cloud computing and the issues involved. In: Proceedings of 2014 11th International Bhurban Conference on Applied Sciences Technology (IBCAST), Islamabad, Pakistan, 14th–18th January 2014, pp. 414–419, January 2014
2. Al Nuaimi, E., Al Neyadi, H., Mohamed, N., Al-Jaroodi, J.: Applications of big data to smart cities. J. Internet Serv. Appl. 6(1), 25 (2015)
3. Arasu, A., Babu, S., Widom, J.: The CQL continuous query language: semantic foundations and query execution. VLDB J. 15(2), 121–142 (2005)
4. Bonino, D., Rizzo, F., Pastrone, C., Soto, J.A.C., Ahlsen, M., Axling, M.: Block-based realtime big-data processing for smart cities. In: 2016 IEEE International of Smart Cities Conference (ISC2), pp. 1–6. IEEE (2016)
5. Carbone, P., Ewen, S., Haridi, S., Katsifodimos, A., Markl, V., Tzoumas, K.: Apache Flink: unified stream and batch processing in a single engine. Data Eng., 28–38 (2015)
6. Cheng, B., Longo, S., Cirillo, F., Bauer, M., Kovacs, E.: Building a big data platform for smart cities: experience and lessons from santander. In: 2015 IEEE International Congress on Big Data (BigData Congress), pp. 592–599. IEEE, June 2015
7. Del Esposte, A.M., Kon, F., Costa, F.M., Lago, N.: InterSCity: a scalable microservice-based open source platform for smart cities. In: Proceedings of the 6th International Conference on Smart Cities and Green ICT Systems (2017)
8. Demchenko, Y., de Laat, C., Membrey, P.: Defining architecture components of the big data ecosystem. In: 2014 International Conference on Collaboration Technologies and Systems (CTS), pp. 104–112, May 2014
9. EsperTech: Complex Event Processing (2014). http://www.espertech.com/esper/
10. Etzion, O., Niblett, P.: Event Processing in Action, 1st edn. Manning Publications Co., Greenwich (2010)
11. Flouris, I., Giatrakos, N., Deligiannakis, A., Garofalakis, M., Kamp, M., Mock, M.: Issues in complex event processing: status and prospects in the Big Data era. J. Syst. Softw. 127, 1–20 (2016)
12. Girtelschmid, S., Steinbauer, M., Kumar, V., Fensel, A., Kotsis, G.: Big data in large scale intelligent smart city installations. In: Proceedings of International Conference on Information Integration and Web-based Applications & Services, p. 428. ACM (2013)
13. Gurgen, L., Gunalp, O., Benazzouz, Y., Gallissot, M.: Self-aware cyber-physical systems and applications in smart buildings and cities. In: 2013 Design, Automation Test in Europe Conference Exhibition (DATE), pp. 1149–1154, March 2013
14. Kon, F., Santana, E.F.Z.: Cidades inteligentes: conceitos, plataformas e desafios. Jornadas de Atualização em Informática 2016—JAI, p. 17 (2016)
15. Luckham, D., Schulte, R.: Event Processing Glossary - Version 2.0 (2011)
16. Luckham, D.C.: The Power of Events: An Introduction to Complex Event Processing in Distributed Enterprise Systems. Addison-Wesley Longman Publishing Co., Inc., Boston (2001)
17. Matysiak, M.: Data stream mining: basic methods and techniques. Technical report, Rheinisch-Westfälische Technische Hochschule Aachen (2012)
18. Mitton, N., Papavassiliou, S., Puliafito, A., Trivedi, K.S.: Combining cloud and sensors in a smart city environment. EURASIP J. Wirel. Commun. Netw. 2012(1), 247 (2012)

19. Parkavi, A., Vetrivelan, N.: A smart citizen information system using Hadoop: a case study. In: 2013 IEEE International Conference on Computational Intelligence and Computing Research, December 2013
20. Sanchez, L., et al.: SmartSantander: IoT experimentation over a smart city testbed. Comput. Netw. **61**, 217–238 (2014)
21. Shasha, D., Bonnet, P.: Database Tuning: Principles, Experiments, and Troubleshooting Techniques. Elsevier, Amsterdam (2002)
22. Tei, K., Gürgen, L.: ClouT: cloud of things for empowering the citizen clout in smart cities. In: 2014 IEEE World Forum on Internet of Things (WF-IoT), pp. 369–370, March 2014
23. Tönjes, R., et al.: Real time IoT stream processing and large-scale data analytics for smart city applications. In: Poster Session, European Conference on Networks and Communications (2014)
24. Yang, J., Leskovec, J.: Patterns of temporal variation in online media. In: Proceedings of the Fourth ACM International Conference on Web Search and Data Mining, WSDM 2011, pp. 177–186. ACM, New York (2011)

SLEDS: A DSL for Data-Centric Storage on Wireless Sensor Networks

Marcos Aurélio Carrero[1,3](✉), Martin A. Musicante[2], Aldri Luiz dos Santos[1], and Carmem S. Hara[1]

[1] Universidade Federal do Paraná, Curitiba, Brazil
{macarrero,aldri,carmem}@inf.ufpr.br
[2] Universidade Federal do Rio Grande do Norte, Natal, Brazil
mam@dimap.ufrn.br
[3] FAE Centro Universitário – Paraná, Curitiba, Brazil

Abstract. The dynamicity requirements of urban sensor networks rise new challenges to the development of data management and storage models. Software component techniques allow developers to build a software system from reusable, existing components sharing a common interface. Moreover, the development of urban sensor networks applications would greatly benefit from the existence of a dedicated programming environment. This paper proposes SLEDS, a Domain-Specific Language for Data-Centric Storage on Wireless Sensor Networks. The language includes high-level composition primitives, to promote a flexible coordination execution flow and interaction between components. We present the language specification as well as a case study of data storage coordination on sensor networks. The current specification of the language generates code for the NS2 simulation environment. The case study shows that the language implements a flexible model, which is general enough to be used on a wide variety of sensor network applications.

Keywords: WSN storage · Software components · Domain-Specific Languages

1 Introduction

Wireless sensor networks (WSNs) are essential components of urban computing. They can be applied in a variety of contexts. For traffic monitoring, they can be used to monitor the flow of vehicles in order to control the traffic lights and minimize jams. For environment monitoring, they can be used to collect the pollution level in order to detect critical areas and take actions that minimize its effect on the population.

Sensor networks deployed on urban areas are usually dense. They are composed of thousands of devices that communicate via radio, and have limited

This research was partially funded by INES 2.0, CNPq grant 465614/2014-0 and Fundação Araucária.

J. Oliveira et al. (Eds.): BiDU 2018, CCIS 926, pp. 74–89, 2019.
https://doi.org/10.1007/978-3-030-11238-7_5

resources for processing and storing data. There are three categories of data storage models for WSNs [13,22]: local, external, and data-centric. The local and external categories store sensed data on the sensor device, and on an external device with more resources (usually called the *base station*), respectively. These categories are not appropriate for dense networks. This is because local storage requires all devices to be contacted in order to collect data to answer queries, which may result in poor response times. On the other hand, external storage requires sensors to periodically report their readings to the base station, which may cause unnecessary high traffic of messages. The data-centric approach, on the other hand, combines both approaches, by electing a subset of sensors to act as representatives of sets of devices, which store their readings. In this model, groups of sensors compose *clusters*, represented by *cluster heads - CHs*. Thus, in order to answer queries, only CHs have to be contacted, providing scalability for dense networks, such as urban WSNs. There are a number of data-centric storage models proposed for urban scenarios [11].

Validation of the proposed models usually involves programming them in a simulation environment, given the costs and difficulties of deploying such large networks in real settings. NS2, NS3, and OMNeT++ are among the simulation environments used for WSNs. While developing previous works on urban data-centric models we have noticed that: *(1)* the majority of programs are developed from scratch, and there is little to no support for code reusability; and *(2)* there are similarities among models on the flow of activities, which can be modeled as state machines. We have tackled these problems by proposing a component-based model for WSNs called RCBM [7], and a state machine to formalize the interaction among components [5]. Although the state machine helps the *specification* of the overall flow of activities, the programmer is still responsible for developing the code specified by the state machine. In this paper, we propose a language that closely resembles a state machine, which allows the programmer to define the flow of activities in a higher-level of abstraction. Our current specification generates code for the NS2 simulator. However, we envision that in the future the same program can be used to generate code for sensor devices using a platform-independent library such as wiselib [4].

The idea of specifying the control flow in a higher-level language can minimize the complexity of developing event-based programs. Event-based programming is often used as an abstraction mechanism for devices with limited resources. In WSNs, this programming model is adopted by operating systems such as TinyOS and Contiki, as well as for simulation environments: NS2, NS3 and OMNeT++. However, the flow of events in these programs is hard to understand and maintain [14]. Examining NS2 programs coded by different developers, we have noticed that they found it difficult to control the flow of activities when they were not triggered by an event, but by a logical condition or a timer. Each programmer used a different approach for handling this type of state change, generating completely different programs, which are thus hard to maintain.

Our proposed language, called SLEDS (State Machine-based Language for Event-Driven Systems), overcomes this problem, by directly defining states and transitions among them. Transitions may be event-based and logic-based. SLEDS also supports primitives for point-to-point and broadcast communication among sensors. The language focus on the coordination flow of data-centric entities that are associated with a set of components that implement common functionalities. In this paper, we present the language specification, as well as a syntax-based translation of SLEDS to NS2. A case study that implements data-centric storage models for WSNs shows that the language is general enough to be used on a wide variety of applications.

The remainder of the paper is organized as follows. Section 2 discusses the related work. Section 3 presents SLEDS as well as a syntax-based translation. Section 4 details a case study of data storage coordination on sensor networks. We conclude in Sect. 5 highlighting future works.

2   Related Work

Data-centric storage has attracted a lot of attention, given that its decentralized approach is more scalable for large-scale urban WSNs than external and local storage models. In this model, some sensors in the network are responsible for storing the readings of a group of sensors. MKSP [9] follows this approach by mapping raw data to storage nodes. In order to exploit the spatial data similarity of sensor readings, AQPM [6] and SILENCE [17] consider some sensors elected as cluster heads to be the group representative, minimizing the communication overhead. Although recent efforts have been made to build efficient data storage systems, the specific nature of WSNs and the lack of a common general purpose development framework make the design of these applications a hard task. RCBM [7] promotes software reuse from existing components to improve the efficiency of system development and evaluation. The separation of the coordinator from the application components proposed by RCBM allows developers to explore similarities among models on the flow of activities.

Domain-Specific Languages (DSL) are programming languages to be used in a well-defined context. As opposed to general-purpose programming languages, DSLs are devised to closely follow practices of their application domain [10]. DSLs are commonly used in the context of Wireless Sensor Networks [8], as well as for the definition of state-transition systems. In the context of WSNs, Hood [23] provides a neighborhood programming abstraction. Algorithms are designed based on a set of criteria for choosing neighbors and definition of variables to share among them. Hood is aimed at simplifying the use of operations such as synchronization and communication with neighboring sensors. SenNet [21] abstracts WSN programming complexity to develop node and group-level applications. Although the purpose of Hood, SenNet and SLEDS is similar, to provide a high-level abstraction for developing sensor-based application, the approaches adopted by each of them differ. SLEDS is based on a state machine while Hood is based on the concept of neighborhood. SenNet does not adopt a flexible execution model such as the one proposed by SLEDS.

Also in the Data-Centric context, Regiment [20] is a DSL which provides a geo-temporal view of the WSN. The language provides primitives to manipulate sets of geo-localized data streams. This centralized view is translated by the compiler into specific code to be run by each sensor in the network. The use of DSLs for the definition of state machines is a well-studied topic [15]. For instance, in [19], the authors propose a DSL to implement a specific type of state machines to describe complex systems. However, it does not target sensor applications as SLEDS. We observed that the development of programs to control data-centric coordination and storage on WSNs follow patterns that can be modeled by state-transition machines. Moreover, developing such applications is usually considered complex by an average programmer. In order to tackle these problems, we defined SLEDS, a DSL for managing data-centric storage on WSNs.

3   SLEDS: The Language

This section presents SLEDS (State Machine-based Language for Event Driven Systems), a DSL used to implement state machines for data storage coordination on sensor networks. Section 3.1 presents the language syntax, followed by a specification of its translation to NS2 network simulation code in Sect. 3.2.

3.1   Syntax

The grammar illustrated in Fig. 1 describes the SLEDS syntax. In our programming model, each sensor executes an instance of a SLEDS program. A state machine communicates with each other through asynchronous message passing. A SLEDS program consists of (*i*) a **Program** declaration, its identifier and a sequence of input parameters, followed by (*ii*) a sequence of constants **const**, (*iii*) a sequence of variables, and (*iv*) a sequence of state definitions **StateDef**.

A **StateDef** is composed of an identifier, a sequence of input parameters followed by a list of actions **ActionList**. The actions correspond to sensor activities triggered in response to an event or based on a logical condition. An **ActionList** is a sequence of standard control flows, such as sequential, conditional, and iteration, as well as primitives for sending and receiving messages, as detailed below:

- Action ::= **nextState** *State*: describes a state change to a new *State*. Each *State* declaration has a name **Id** with arguments representing the input parameters of the state. The *exit* state finishes the program.
- Action ::= **broadcast** (*Exp, Exp, ExpList*): corresponds to the asynchronous communication sent from a sensor to all its neighbor sensors, that is, the ones within its communication range. The arguments are the message type, message identifier and a list of parameters.
- Action ::= **send** (*Exp, Exp, ExpList, ExpList*): corresponds to the asynchronous communication sent from a sensor to a set of destination sensors. The arguments are the message type, message identifier, the set of destinations and a list of parameters.
- Action ::= **on recvBroadcast** (*Id, Id, IdList*){ *ActionList* }: corresponds to the receipt of a *broadcast* message.

```
Program      ::= Program Id(Type Id(, Type Id)*) {
                 (const Id = (Num-Literal | Str-Literal); )*
                 (Type VarList; )* StateDef* }
VarList      ::= Var (, Var)*;
Var          ::= Assignment | Id
StateDef     ::= State Id(Type Id(,Type Id)*) { ActionList }
State        ::= Id ( ExpList? ) | exit
ActionList   ::= Action (Action)*
Action       ::= nextState State;
               | broadcast(Exp, Exp, ExpList);
               | send(Exp, Exp, ExpList, ExpList);
               | on recvBroadcast(Id, Id, IdList) {ActionList}
               | on recv(Id, Id, IdList, IdList) { ActionList }
               | during(Exp) on recvBroadcast(Id, Id, IdList)
                 { ActionList } nextState State;
               | during (Exp) on recv(Id, Id, IdList, IdList)
                 { ActionList } nextState State;
               | while (Exp) { ActionList }
               | for Id in Exp { ActionList }
               | if (Exp) { ActionList } (else { ActionList })?
               | Assignment;
               | Method-call;
Method-call  ::= Exp
Assignment   ::= Id = Exp
Exp          ::= Exp − > Exp | Exp . Id | Exp(Exp?) | Id
ExpList      ::= Exp (, Exp)*
IdList       ::= Id (, Id)*
```

Fig. 1. SLEDS syntax

- Action ::= **on recv** (*Id, Id, IdList, IdList*){ *ActionList* }: corresponds to the receipt of a *send* message.
- Action ::= **during** (*Exp*) **on recvBroadcast** (*Id, Id, IdList*) { *ActionList*} **nextState** *State*: corresponds to the receipt of a *broadcast* message during a time interval, and change to a new *State* at the end of this period.
- Action ::= **during** (*Exp*) **on recv** (*Id, Id, IdList, IdList*) { *ActionList* } **nextState** *State*: corresponds to the receipt of a *send* message during a time interval, and change to a new *State* at the end of this period.

As an example, consider the state machine for discovering neighbors illustrated in Fig. 2. Note that there are two types of transitions:

- event state change: specifies a transition to a new state when the sensor receives a message or upon the timer expiration (represented in blue lines).
- logic state change: a machine transitions to a new state triggered by the result of a computation, and represented in red lines.

Some event-based languages, such as NS2 simulator, provide limited abstraction to implement state machine models. The SLEDS language facilitates this

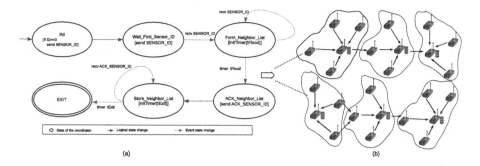

Fig. 2. State machine of a discovery neighbors algorithm. (Color figure online)

task allowing developers to describe the state machine in a high level abstraction and translates this representation into an NS2 simulation code. Listing 1.1 illustrates the SLEDS program that implements the neighbor discovery machine.

```
1   use compSensor as ComponentsSensor;
2   use compLibMSG as ComponentsLibMessage;
3
4   Program Coordinator() {
5     const tFlood=25;
6     const tExit=0.1;
7     int myID = compSensor->getSensorId();
8     list<int> listSensorAnnouncements;
9     int msgID;
10
11    STATE INI() {
12      if (myID == 0) {
13        msgID = compLibMSG->GetNextMsgId();
14        broadcast(SENSOR_ID, msgID, myID);
15        compLibMSG->addSeenMsg(SENSOR_ID, msgID); }
16      nextState Wait_First_Sensor_ID(); }
17
18    STATE Wait_First_Sensor_ID() {
19      on recvBroadcast(SENSOR_ID, msgID, ID) {
20        listSensorAnnouncements.insert(ID);
21        if (!compLibMSG->seenMsg(SENSOR_ID, msgID)) {
22          compLibMSG->addSeenMsg(SENSOR_ID, msgID);
23          broadcast(SENSOR_ID, msgID, myID); }
24        nextState Form_Neighbor_List();   } }
25
26    STATE Form_Neighbor_List() {
27      during (tFlood) on recvBroadcast(SENSOR_ID, msgID, ID) {
28        listSensorAnnouncements.insert(ID); }
29      nextState ACK_Neighbor_List(); }
30
31    STATE ACK_Neighbor_List() {
32      send(ACK_SENSOR_ID, compLibMSG->GetNextMsgId(),
33           listSensorAnnouncements, myID);
34      nextState Store_Neighbor_List(); }
35
36    STATE Store_Neighbor_List() {
37      during (tExit) on recv(ACK_SENSOR_ID, msgID,
38                       listSensorAnnouncements, fromID) {
39        for v in listSensorAnnouncements
40          if (v == myID)
41            compSensor->listKnownNeighbors.insert(fromID); }
42      nextState exit;}
```

Listing 1.1. Neighbor discovery SLEDS program

The program assumes the existence of components: *compSensor* and *compLibMSG*. The first provides basic sensor functionality, such as returning the sensor identification (function *getSensorId*), and storage of its list of neighbors (*listKnownNeighbors*). The *compLibMSG* provides functionality related to messages exchanged among sensors. There are functions to create a new message identification (*GetNextMsgId*) and to store, for each sensor, the type and identification of messages already received (*addSeenMsg*). In fact, in a component-based programming environment, a SLEDS program plays the role of a coordinator, which is responsible for the flow of activities that glue the software components, specifying the interactions among them. After including references to the components *(l. 1–2)*, the program declares a set of constants and variables. Constant *tFlood* defines the delay of message transmissions in the network, and constant *tExit* determines the delay needed between sending a message and receiving an acknowledgement in order to avoid collisions. Variable *myID* keeps the sensor unique identifier, which is obtained executing function *getSensorId()* provided by the *compSensor* component *(l. 7)*.

Every sensor executes the same SLEDS program, starting in the INI state *(l. 11)*. In this state, the sensor with *myID* zero, obtains a new message identifier and sends a message of type SENSOR_ID to all its neighbors *(l. 12–14)*, and stores the message identifier type and identifier in order to avoid sending duplicated messages *(l. 15)*. The sensor with *myID* zero and the remaining sensors perform a logical state change to Wait_First_Sensor_ID *(l. 16)*. The sensors that receive the first message SENSOR_ID store the *ID* contained in the message and send their identifier to their neighbors *(l. 19–23)* and perform an event state transition to Form_Neighbor_List *(l. 24)*. In state Form_Neighbor_List, sensors continue to store the ID from their neighbors during a time interval *tFlood (l. 26–28)*. When the timer expires, nodes perform an event state change to ACK_Neighbor_List *(l. 29)* and send an ACK message to the known neighbors recorded in variable *listSensorAnnouncements (l. 32)*. After sending the ACK message, the sensors make a logical state change to Store_Neighbor_List *(l. 34)*. In state Store_Neighbor_List, during a time interval *tExit (l. 36–38)*, sensors that receive the message check if they are the final destination and update its list of neighbors *(l. 40–41)*. At the end of the flooding, each sensor has in its local *listKnowNeighbors* variable, its list of neighbors. Next section presents a proposal to translate SLEDS programs to NS2 simulation codes.

3.2   Translation to NS2

In a NS2 program, the coordination of the sensor activities is implemented in two main functions: recv and TimerHandle. Function recv is responsible for managing messages and contains the code for state transitions triggered upon a message receipt. Function TimerHandle is activated by the expiration of a timer. Observe that in the state machine illustrated in Fig. 2, both are represented as event-based transitions, and there is no distinction among them in the SLEDS program in Listing 1.1. However, in the NS2 program, they have to be coded in different functions, which adds complexity to understand the flow of activities. Moreover, states may generate code not only for the `recv` or `TimerHandle` functions, but for both. Examples for each case are presented next.

```
1  STATE INI() {
2    if (myID == 0) {
3      msgID = compLibMSG→GetNextMsgId();
4      broadcast(SENSOR_ID, msgID, myID);
5      compLibMSG→addSeenMsg(SENSOR_ID, msgID);
6    }
7    nextState Wait_First_Sensor_ID();
8  }
```

```
1  void WSN_ComponentsAgent::TimerHandle(State st) {
2    switch (st) {
3      case INI: {
4        if (myID == 0) {
5          msgID = compLibMSG→GetNextMsgId();
6          broadcast(SENSOR_ID, msgID, myID);
7          compLibMSG→addSeenMsg(SENSOR_ID, msgID);
8        }
9        nextState=Wait_First_Sensor_ID;
10      }
11    }
12  }
```

Fig. 3. SLEDS code and NS2 translation of INI state

The INI state of Listing 1.1 is an example that generates code only for the `TimerHandle` function as shown in Fig. 3. The translated code is composed of a *switch* command, with *case* clauses, one for each state identifier. The *case* for the INI state contains the same code provided by the SLEDS program, which ends with a state transition to state `Wait_First_Sensor_ID` (*l.9*). The translation of this state generates code only for the `recv` function, as illustrated in Fig. 4.

```
1  STATE Wait_First_Sensor_ID() {
2    on recvBroadcast(SENSOR_ID, msgID, ID) {
3      listSensorAnnouncements.insert(ID);
4      if (!compLibMSG→seenMsg(SENSOR_ID, msgID)) {
5        compLibMSG→addSeenMsg(SENSOR_ID, msgID);
6        broadcast(SENSOR_ID, msgID, myID);
7      }
8      nextState Form_Neighbor_List();
9    }
10 }
```

```
1  void WSN_ComponentsAgent::recv(Packet* pkt, Handler *) {
2    WSN_Components_Message p = pkt;
3    switch(nextState) {
4      case (Wait_First_Sensor_ID): {
5        if (param.getMsgType() == SENSOR_ID) {
6          listSensorAnnouncements.insert(param.getSensorID());
7          if (!compLibMSG→seenMsg(p.getSensorID(), p.getMsgID())) {
8            compLibMSG→addSeenMsg(p.getSensorID(), p.getMsgID());
9            broadcast(SENSOR_ID, msgID, myID);
10          }
11          nextState=Form_Neighbor_List;
12        }
13      }
14    }
15 }
```

Fig. 4. SLEDS code and NS2 translation of Wait_First_Sensor_ID state

Similar to `TimerHandle`, the `recv` function in NS2 is also composed of a *switch* command, with *case* clauses, one for each state. Part of the generated code in NS2 is to obtain the parameters from the packet received, but most of the code inside each *case* clause is identical to the SLEDS program. The translation to NS2 is not so direct when it involves both functions `TimerHandle` and `recv`, as shown in Fig. 5, which corresponds to state `Form_Neighbor_List`. The `recv` function sets a timer `tFlood` (*l.7*). During this period the sensor stores the neighbor announcements (*l.10*) at every `SENSOR_ID` message received. However, the transition to the next state `ACK_Neighbor_List` cannot be made in this function, since the sensor may receive multiple message of this type. Thus, the transition is coded in the `TimerHandle` function, which is triggered at the expiration of `tFlood` timer.

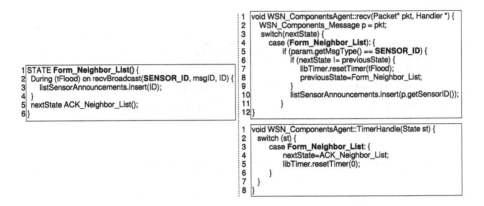

```
1 STATE Form_Neighbor_List() {
2   During (tflood) on recvBroadcast(SENSOR_ID, msgID, ID) {
3     listSensorAnnouncements.insert(ID);
4   }
5   nextState ACK_Neighbor_List();
6 }
```

```
1  void WSN_ComponentsAgent::recv(Packet* pkt, Handler *) {
2    WSN_Components_Message p = pkt;
3    switch(nextState) {
4      case (Form_Neighbor_List): {
5        if (param.getMsgType() == SENSOR_ID) {
6          if (nextState != previousState) {
7            libTimer.resetTimer(tFlood);
8            previousState=Form_Neighbor_List;
9          }
10         listSensorAnnouncements.insert(p.getSensorID());
11       }
12 }
```

```
1  void WSN_ComponentsAgent::TimerHandle(State st) {
2    switch (st) {
3      case Form_Neighbor_List: {
4        nextState=ACK_Neighbor_List;
5        libTimer.resetTimer(0);
6      }
7    }
8 }
```

Fig. 5. SLEDS code and NS2 translation of Form_Neighbor_List state

Due to space limit, we have not included the translation for states `ACK_Neighbor_List`, which generates code in function `TimerHandle`, and `Store_Neighbor_List`, which generates code in both functions.

We have adopted a syntax directed translation to generate the NS2 code from a SLEDS program. In this technique, *attributes* and *semantic rules* are associated to each production of the grammar [1]. The general approach consists of building a derivation tree and then determine the attribute values in each node of the tree during its traversal. Semantic rules express the relationship between the computation of attribute values and the productions, by associating code fragments to each attribute. The set of attributes and semantic rules is denoted as an *attribute grammar* [18]. There are two types of attributes: synthesized and inherited. In our approach, synthesized attributes are used to pass semantic information up the parse tree, while inherited attributes help pass semantic information down.

In our grammar, there are three attributes: `rc`, `tc`, and `dest`. The first two are synthesized attributes and determine whether the code fragment is going to be included in function `recv` or `TimerHandle`, respectively. Attribute `dest` is inherited, and may contain either the value `rc` or `tc`, and it is used to pass down the tree the function in which the code will be generated, based on the node context.

The attribute grammar in Figure 6, describes the attributes and grammar rules to generate code for the state `Form_Neighbor_List`, and Fig. 7 the resulting parse tree. Observe at the bottom of the tree that the inherited attribute `dest` of node `Action` contains the value `rc` in order to pass the information down the tree that the `Method-call` should generate code for attribute `rc`. This attribute, will then receive the code fragment, which is passed up the tree in order to compose

(r1)	$State ::=$ **State$_1$** Id_1 $("Type\ Id_2(,Type\ Id_3)^*)$ "{" $ActionList$ "}" $\quad\|$ if $(ActionList.tc\ \mathtt{!=\ null})$ $\qquad State.tc =$ "switch (nextState):{ case" $+ Id_1.txt +$ ":" $\qquad\qquad +$ "{" $ActionList.tc +$ "}}" $\quad\|$ if $(ActionList.rc\ \mathtt{!=\ null})$ $\qquad State.rc =$ "switch (nextState):{ case" $+ Id_1.txt +$ ":" $\qquad\qquad +$ "{"$+ ActionList.rc +$ "}}"
(r2)	$Action ::=$ **during** $("Exp")$ **on recvBroadcast** $("Id_1, Id_2, IdList")$ $\qquad$ "{" $ActionList$ "}" $nextState\ State")$ $\quad\|\ Action.rc =$ "if (nextState != previousState) {" $\quad\ Action.rc +=$ "libTimer.resetTimer("$+Exp.txt+$")" $\quad\ Action.rc +=$ "previousState="$+Action.pst$ $\quad\ Action.rc +=\ ActionList.rc\ +$ "}"
(r3)	$Method - call ::=\ Exp$ $\quad\|\ \mathtt{if}(Method - call.dest\mathtt{==rc})\quad \{Method - call.rc\ =\ Exp.rc\}$ $\quad\ \mathtt{else}\ \{\ Method - call.tc\ =\ Exp.tc\ \}$

Fig. 6. Attribute grammar to generate code for state Form_Neighbor_List

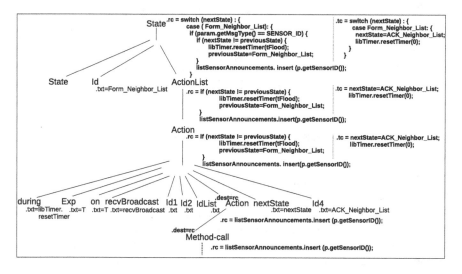

Fig. 7. Parse tree for the Form_Neighbor_List state

the final `rc` value at the tree root. This is the code that will be included in the `recv` function. The same process is used to compose the final value of the `tc` attribute, with the code to be included in the `TimerHandle` function.

Examples presented in this section show how a SLEDS program can generate NS2 code using an attribute grammar specification. The next section presents a case study that shows how SLEDS programs can be used to generate data-centric storage models for WSNs in a component-based development framework.

4 Validation

RCBM [7] is a component-based framework to develop WSNs storage models that promotes code reusability. It is depicted in the central box of Fig. 8. RCBM addresses data-centric entities that share concepts and functionalities, which represent various instances of WSN storage systems. These shared functionalities are the components of the system. Although it has been shown that the framework is efficient for promoting code reusability, the application developer is still responsible for coding the coordination among the components. SLEDS, with its high-level composition primitives, can be used to generate code for the RCBM coordinator, as we will show in this section.

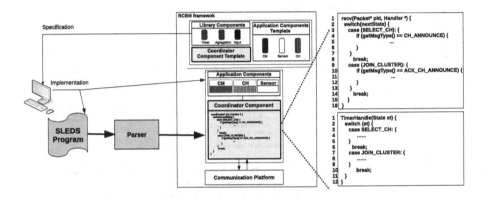

Fig. 8. SLEDS back-end architecture.

RCBM has been implemented on the NS2 simulator and considers three types of components: library components, application components, and the coordinator. Library components provide a toolbox, that can be used to implement application components, associated with WSN entities. For data-centric models, these entities include: *(i)* sensor devices; *(ii)* cluster members (CM), which consist of a set of sensors; and *(iii)* cluster-heads (CHs) that are sensors responsible for storing the information of all cluster members. These entities define a hierarchical storage model, where each cluster designates a sensor as cluster-head for storing the readings of its group members. The coordinator is responsible for the execution flow and message exchanges. Next section shows a case study that implements data-centric storage models for WSNs.

4.1 LEACH Coordinator Component Implementation

LEACH (Low-Energy Adaptive Clustering Hierarchy) [16] is a probabilistic model that forms one-hop clusters. LEACH assumes that all nodes are within the communication range of each other. Sensors elect themselves as cluster-heads with a probability p. In RCBM, the compCH component defines the function selectCH(map<K, V>) that the developer should implement according to the target model. For the coordination that implements the cluster formation of LEACH, every sensor s_i executes selectCH(s_i, p), where $K = s_i$ and $V = p$. Listing 1.2 depicts the SLEDS coordination code of the CH election phase.

```
1    // Program executed by each sensor
2    use compSensor as WSN_ComponentsSensor;
3    use compCH as    WSN_ComponentsCH;
4    use compCM as    WSN_ComponentsCM;
5    use compLibMSG as ComponentsLibMessage;
6
7    Program Coordinator() {
8      const p=0.2;
9      const tCluster=25;
10     const tExit=0.1;
11
12     double RSS;
13     int myCH;
14     int myID=compSensor->getSensorId();
15     list<int, double> knownCHs;
16     list<int> sensorList;
17
18     STATE Select_CH() {
19     if (compCH->selectCH(myID, p)) {
20        broadcast(CH_ANNOUNCE, compLibMSG->GetNextMsgId(), myID);
21        compSensor->role = CH; }
22     else {
23        compSensor->role = CM; }
24     nextState Join_Cluster(); }
```

Listing 1.2. The CH Election Coordination

The Select_CH state (*l.18*) describes the actions that should be executed during the election phase. First, each node calls selectCH() (*l.19*). If the function returns true then the sensor broadcasts its role as cluster-head (CH) to the network (*l.20–21*) and performs a state transition to Join_Cluster (*l.24*). Otherwise, the sensor role is set as a cluster member (CM) (*l.22–23*). Listing 1.3 illustrates the Join_Cluster state code.

```
25   STATE Join_Cluster() {
26   During (tCluster) on recvBroadcast(CH_ANNOUNCE, msgID, ID) {
27     RSS = compSensor->getRSS(ID);
28     knownCHs.insert(ID, RSS); }
29   nextState Cluster_Formation(); }
```

Listing 1.3. The Join Cluster Coordination

In the next state (*l.25*), sensor nodes wait for tCluster time units for CH announcements and update the value of knownCHs based on the received signal strength (RSS) (*l.26–28*). When the timer expires, remaining sensors join the cluster, as illustrated in Listing 1.4.

```
30   STATE Cluster_Formation() {
31     if (compSensor->role = CM) {
32        myCH = compCM->joinCluster(knownCHs);
33        sensorList.insert( myCH );
34        send(ACK_CH_ANNOUNCE, compLibMSG->GetNextMsgId(), sensorList, myID);
35        nextState EXIT;
36     } else {
37        nextState Store_Members(); }
```

Listing 1.4. The Cluster Formation Coordination

In LEACH, a cluster member (*l.31*) decides to join the cluster that requires the lowest energy consumption to communicate. Thus, it sets as CH the sensor with the maximum RSS recorded by knownCHs (*l.32*). Then, it sends an ACK_CH_ANNOUNCE message to the chosen one and move to the EXIT state (*l.33–35*). Otherwise, the CHs perform a logical state change to Store_Members (*l.36–37*). Listing 1.5 depicts the SLEDS code implementation.

```
38   STATE Store_Members() {
39     During (tExit) on recv(ACK_CH_ANNOUNCE, msgID, destListID, fromID) {
40       for v in destListID{
41         if (v = myID) {
42           compCH->members.insert( fromID ); } } }
43     nextState EXIT ;}
```

Listing 1.5. The Store Members Coordination

The CHs execute the actions corresponding to the Store_Members state (*l.38*). First, CHs wait for timer tExit units to receive ACK_CH_ANNOUNCE messages from group members (*l.39*). If it receives a message, the sensor updates its list structure of members (*l.40–42*). When the timer tExit expires, the program terminates its execution (*l.43*). Figure 9 illustrates a state machine of the LEACH coordination flow, an instance of a data-centric WSN storage system.

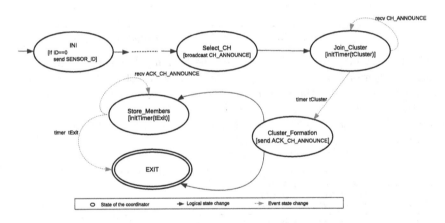

Fig. 9. State machine of a data-centric storage model.

The flow of execution depicted in Fig. 9 is similar to the one adopted by the component-based framework CBCWSN [2], which has been shown to express a number of data-centric storage instances. As we will show next, the same can be said of the SLEDS program. In order to implement LCA [3], only three lines of code have to be modified, mainly to take into consideration a distinct criterion for CH election. LCA elects as CH the sensor with the lowest ID among its neighbors that not received a CH announcement. Listing 1.6 shows the two states in which there are lines in the SLEDS program that differ from the LEACH code. Line 27 from Listing 1.3 has been removed and Lines 5 and 14 differ on the arguments to functions selectCH and knownCHs.insert.

```
1   Program Coordinator() {
2     list<int> knownNeighbors;
3
4     STATE Select_CH() {
5     if (compCH->selectCH(myID, knownNeighbors)) {
6       broadcast(CH_ANNOUNCE, GetNextMsgId(), myID);
7       compSensor->role = CH; }
8     else {
9       compSensor->role = CM; }
10    nextState Join_Cluster(); }
11
12    STATE Join_Cluster() {
13    During (tCluster) on recvBroadcast(CH_ANNOUNCE, msgID, ID) {
14      knownCHs.insert(ID); }
15    nextState Cluster_Formation(); }
```

Listing 1.6. The LCA Coordination

The two case studies presented in this section show that the model instances share the same state machine specification, promoting reusability. The programmer develops a few lines of code with the specificities of each model. Moreover, the state machine primitives adopted by SLEDS does not impose any fixed flow of activities (such as CBCWSN), but allow the developer to define the coordination of any data-centric model.

5 Conclusion

In this paper we proposed a Domain-Specific Language, called SLEDS, for prototyping Wireless Sensor Network applications that adopt a data-centric storage approach. The current specification of the language generates code to run on the NS2 simulation environment, using a library of components provided by RCBM [7]. We validate our approach by defining a syntax-directed translation into NS2 code. As case studies we developed SLEDS programs for LEACH and LCA data-centric models. Our experiments showed that both models share many similarities on the flow of activities. The achieved results show that SLEDS allowed code reuse and agile development for the LCA specification. Our proposal answers some of the challenges identified in [12]. In the future, we intend to implement the parser to translate SLEDS program to NS3 code, a more intuitive NS2 evolution as well as to other simulators and real networks.

References

1. Aho, A.V., Lam, M.S., Sethi, R., Ullman, J.D.: Compilers: Principles, Techniques, and Tools, 2nd edn. Addison-Wesley Longman Publishing Co., Inc., Boston (2006)
2. Amaxilatis, D., Chatzigiannakis, I., Koninis, C., Pyrgelis, A.: Component based clustering in wireless sensor networks. arXiv preprint arXiv:1105.3864 (2011)
3. Baker, D.J., Ephremides, A.: A distributed algorithm for organizing mobile radio telecommunication networks. In: Proceedings of the 2nd International Conference on Distributed Computing Systems, Paris, France, pp. 476–483 (1981)

4. Baumgartner, T., Chatzigiannakis, I., Fekete, S., Koninis, C., Kröller, A., Pyrgelis, A.: Wiselib: a generic algorithm library for heterogeneous sensor networks. In: Silva, J.S., Krishnamachari, B., Boavida, F. (eds.) EWSN 2010. LNCS, vol. 5970, pp. 162–177. Springer, Heidelberg (2010). https://doi.org/10.1007/978-3-642-11917-0_11
5. Carrero, M., Zamproni, K., Musicante, M.A., Santos, A., Hara, C.: Uma máquina de estados para especificação de códigos de simulação para redes de sensores sem fio urbanas. In: Simpósio Brasileiro de Computação Ubíqua e Pervasiva (2018)
6. Carrero, M.A., da Silva, R.I., dos Santos, A.L., Hara, C.S.: An autonomic in-network query processing for urban sensor networks. In: 20th IEEE Symposium on Computers and Communications (ISCC), pp. 968–973 (2015)
7. Carrero, M.A., Musicante, M.A., dos Santos, A.L., Hara, C.S.: A reusable component-based model for WSN storage simulation. In: Proceedings of the 13th ACM Symposium on QoS and Security for Wireless and Mobile Networks, pp. 31–38 (2017)
8. Chandra, T.B., Dwivedi, A.K.: Programming languages for wireless sensor networks: a comparative study. In: 2015 2nd International Conference on Computing for Sustainable Global Development (INDIACom), pp. 1702–1708, March 2015
9. D'Angelo, G., Diodati, D., Navarra, A., Pinotti, C.M.: The minimum k-storage problem: complexity, approximation, and experimental analysis. IEEE Trans. Mob. Comput. 15(7), 1797–1811 (2016)
10. van Deursen, A., Klint, P., Visser, J.: Domain-specific languages: an annotated bibliography. SIGPLAN Not. 35(6), 26–36 (2000)
11. Diallo, O., Rodrigues, J.J.P.C., Sene, M., Mauri, J.L.: Distributed database management techniques for wireless sensor networks. IEEE Trans. Parallel Distrib. Syst. 26(2), 604–620 (2015)
12. Estrin, D., Govindan, R., Heidemann, J.S., Kumar, S.: Next century challenges: scalable coordination in sensor networks. In: Kodesh, H., Bahl, V., Imielinski, T., Steenstrup, M. (eds.) MobiCom, pp. 263–270. ACM (1999)
13. Fangfang, L., Zhibo, F., Chuanwen, L., Jia, X., Ge, Y., Shenyang, C.: A data storage method based on multilevel mapping index in wireless sensor networks. In: International Conference on Wireless Communications, Networking and Mobile Computing, pp. 2747–2750 (2007)
14. Fischer, J., Majumdar, R., Millstein, T.: Tasks: language support for event-driven programming. In: Proceedings of the 2007 ACM SIGPLAN Symposium on Partial Evaluation and Semantics-Based Program Manipulation, pp. 134–143. ACM (2007)
15. Fowler, M.: Domain Specific Languages, 1st edn. Addison-Wesley Professional, Boston (2010)
16. Heinzelman, W.R., Chandrakasan, A., Balakrishnan, H.: Energy-efficient communication protocol for wireless microsensor networks. In: 33rd Annual Hawaii International Conference on System Sciences (HICSS-33), 4–7 January 2000, Maui, Hawaii, USA (2000)
17. Lee, E.K., Viswanathan, H., Pompili, D.: Distributed data-centric adaptive sampling for cyber-physical systems. TAAS 9(4), 21:1–21:27 (2015)
18. Louden, K.C.: Compiler Construction: Principles and Practice. PWS Publishing Co., Boston (1997)
19. Murr, F., Mauerer, W.: MCFSM: globally taming complex systems. In: Proceedings of the 3rd International Workshop on Software Engineering for Smart Cyber-Physical Systems. SEsCPS 2017, pp. 26–29. IEEE Press, Piscataway (2017)

20. Newton, R., Morrisett, G., Welsh, M.: The regiment macroprogramming system. In: 2007 6th International Symposium on Information Processing in Sensor Networks, pp. 489–498, April 2007

21. Salman, A.J., Al-Yasiri, A.: Developing domain-specific language for wireless sensor network application development. In: 11th International Conference for Internet Technology and Secured Transactions, ICITST 2016, pp. 301–308 (2016)

22. Shen, H., Zhao, L., Li, Z.: A distributed spatial-temporal similarity data storage scheme in wireless sensor networks. IEEE Trans. Mob. Comput. **10**(7), 982–996 (2011)

23. Whitehouse, K., Sharp, C., Brewer, E., Culler, D.: Hood: a neighborhood abstraction for sensor networks. In: Proceedings of the 2nd International Conference on Mobile Systems, Applications, and Services, MobiSys 2004, pp. 99–110. ACM (2004)

Extraction and Exploration of Business Categories Signatures

Leonardo de Assis da Silva and Thiago H. Silva$^{(\boxtimes)}$

Department of Informatics, Federal University of Technology - Paraná,
Curitiba, PR, Brazil
`leosil@alunos.utfpr.edu.br`, `thiagoh@utfpr.edu.br`

Abstract. Different business types may have distinct businesses functioning dynamics, i.e., popularity times, that can be dictated not only by the service offered but also due to other aspects. Performing the business popularity time comprehension allows us, for instance, to use this information as a business descriptor that could be explored in new services. Recently, Google launched a service, namely Popular Times, which provides the popularity times of commercial establishments. In this study, we collected and analyzed a large-scale dataset provided by that service for business in different cities in Brazil and in the United States. Our main contributions are: (1) clustering and analysis of the collected business popularity times dataset in each studied city; (2) approach for identifying the signature that represents the behavior of specific categories of venues; (3) training and evaluation of an inference model for categories of establishments; (4) user evaluation of some of our results.

Keywords: Google Popular Times · Time series · Signature ·
Large scale urban assessment

1 Introduction

Urban computing is a field of study that, among others objectives, aims help to understand urban phenomenon envisioning to offer smarter urban services. There are several phenomena worth investigating in the city, for instance, citizens mobility, by any transportation mode, citizens interaction, for example, through phone calls, and businesses functioning dynamics, i.e., their popularity times [13,21].

Different types of business may have distinct popularity times that can be dictated not only by the service offered but also due to diverse economic, social and cultural aspects, such as typical working times and nap periods that might exist in certain cities. Besides, businesses in the same category may also eventually exhibit different peaks of popularity according to their location or particular characteristics of the business. While some types of restaurants might be more popular at night, fast-food venues can present a more evenly distributed popularity throughout the day.

© Springer Nature Switzerland AG 2019
J. Oliveira et al. (Eds.): BiDU 2018, CCIS 926, pp. 90–104, 2019.
https://doi.org/10.1007/978-3-030-11238-7_6

Uncover the functioning dynamics, i.e., the popularity times, of business venues is not a trivial task. However, recently, Google launched a service, namely Popular Times, which provides the popularity times for those type of places. This is possible, among other factors, due to the high number of users who use Google's location-based mobile services, enabling Google to know when users visit a certain venue [4].

In this study, we collected and analyzed a large-scale dataset from Google Popular Times for business related to drinking and food consumption habits in different cities in Brazil and the United States. To the best of our knowledge, the present work is the first to explore this source of information. Also, we study patterns of popularity times considering different cities of the same country. Knowing the different patterns associated with specific business locations, e.g., countries, cities, or neighborhoods, could help to improve the description of the functioning of the business and in the understanding of how people from different geographical areas interpret the usefulness and the use of each business category.

That information could also be used in conjunction with business recommendation algorithms to make them sensitive to the local context, e.g., average less busy times. It could also be explored in a market study, for instance, to open new branches in other countries by studying cultural differences, such as the use of spaces. I.e., while a Brazilian can understand that the primary use of restaurants is for lunch, a user from the United States could assume that dinner is the preferred time to visit these types of place.

The main contributions of this work are: (1) grouping and analysis of popularity time series of business venues. We find, for instance, patterns that are related to local factors coming from cities of the same country; (2) approach to identify the signature that represents the popularity times for categories of business. We show that it was possible to find an association between different signatures of categories that can be used for the proposition of meta-categories of business places, which could be more informative about venues; (3) training and evaluation of a classification model to infer the category of a new establishment given its popularity; (4) user evaluation of some of our results, particularly, the meta-categories of business places proposed in this study.

The remainder of this study is divided as follows. Section 2 presents the related works. Section 3 shows the description of our collected dataset and how it is processed. Section 4 explains the clustering process and the procedure for generating category signatures. Section 5 discusses the results, including a possible application of category signatures in the category inference task given a popularity time series. Section 6 discusses an experiment with users to help to validate the proposed meta-categories. Finally, Sect. 7 presents the final considerations and future work.

2 Related Works

Recently, data extracted from the Web has been explored to help in the better understanding of the urban social behavior and city dynamics. For instance, check-ins of Foursquare were used to identify the functional use of city areas, e.g., a shopping area [17]. Areas of points of interest, such as sights and popular establishments, were identified from shared photos on Instagram [14]. In addition, check-ins were used to better understand the state of the traffic in urban regions in [16]. Thus, we highlight three groups of studies that are relevant to the present work: distribution of popular elements, clustering of time series and generation of a representative element of the cluster, and semantic extraction of geolocalized data.

Check-ins shared on Foursquare were used to investigate the properties of Location-Based Social Networks (LBSN); one of the discoveries is the presence of a power law distribution of the popularity of establishments, that is, a small number of establishments receive a high number of visits, while most establishments have low popularity. Consequently, the analysis of a limited sample of popular establishments is still capable of revealing the behavior of a significant proportion of visitors to a category of establishment. This finding is relevant here as the Google Popular Times service only offers information for establishments above a certain threshold of popularity [4]. Knowing that, Neves et al. [9] evaluated the possibility of reproducing Google Popular Times using data extracted from Foursquare to the cities of Curitiba and Chicago. They found evidence that Google Popular Times is consistent with data voluntarily actively shared through check-ins on Foursquare. That indicates that the reproduction of Popular Times for places that do not have this information might be possible using alternative data sources.

Time series clustering has been applied to find patterns across domains through distinct approaches. The task of identifying clusters of time series requires the proposition of new approaches or the use of conventional algorithms together to a suitable distance measure [8]. The variation patterns of *memes* mention in Twitter messages was observed using a new algorithm specially developed to handle temporal data, called K-Spectral Centroid (K-SC), which is an adaptation of the K-means algorithm [18]. Next, the K-SC algorithm was also explored to understand YouTube videos popularity [3]. The criterion used to define the number of clusters in both studies was the Silhueta method and the Hartigan index, both cluster validity indices that evaluates the similarity between elements of the same cluster compared to the similarity in relation to elements of the other clusters [1].

Regarding the application of conventional algorithms and the need for adequate distance measurements, the Dynamic Time Warping (DTW) measure has obtained satisfactory results in several standard datasets [10]. DTW is a dynamic programming technique similar to the editing distance, or distance of Levenshtein, that seeks to find the optimal global alignment between two time series. The DTW is particularly interesting in the context of discovering the shape of popularity signatures because it helps to find similar time series even when offsets occur, whereas this could eventually not happen when performing the comparison through a point to point distance measure. However, due to its time complexity, the DTW distance usually is only applied on short time series or small datasets [12]. To minimize this limitation Petitjean *et al.* proposed a heuristic for the calculation of averages called *DTW Barycenter Averaging* (DBA).

The extraction of knowledge from temporal data can be performed based on information from different domains. For example, the inference task of describing a venue, such as its category, can be accomplished by examining the distribution of the number of check-ins in establishments by hour and by day of the week. Such characteristics were used in conjunction with the spatial location and information of each visitor, such as age and gender, to assign a label to locations through a binary support vector machine trained with data from LBSN in [19] and boosted decision trees trained with data collected through surveys in [7].

This work uses the K-means partitioning algorithm, instead of adapting conventional algorithms as in [18], but using the same clustering validation measures of the studies mentioned above to help in the definition of the number of clusters. Different from the other category inference studies, we obtained the data used here from the Google Popular Times, where the sensing process using the user's device occurs through opportunistic sensing, that is, this source is not dependent on the initiative of users. As it is a background service, it has the potential to be less affected by factors such as a desire to omit visits to certain facilities than LBSNs. In addition, as far as we know, we present the first study to discover and explore business categories popularity signature in different cities.

3   Data Collection and Procedures

In the Google Popular Times service, the distribution of hourly visits, i.e., popularity time, for each day of the week represent the average number of visits to the establishment over several weeks. This information is generated from data sent anonymously by people who have agreed to participate in the Google History Location service by tracking automatic positioning of the device over time via GPS, WI-FI, and mobile network [4].

To collect data from Google Popular Times we explored a web crawler fed with a list of establishments containing the attributes: name, category, city, and country. We created this list with the help of the Yelp Developers API [20]. Explore this source is interesting because it provides information on establishments in a standard format independent of the country. Another strategy would explore open data, which has been increasingly provided by cities. However, the availability of open data is dependent on local policies, so that the scalability of the number of cities and countries could be impaired.

To investigate how people from different geographical areas interpret each category of establishment, we study cities in Brazil (Curitiba, Rio de Janeiro, and São Paulo), and in the United States (Chicago, New York, and San Francisco). One of our goals is to identify the patterns of popularity times exhibited by establishments related to the consumption of food and drink. Specifically, we analyze the following categories from the reference Yelp platform: bakery, bar, coffee, dance club (nightclubs), and restaurant. Look at these type of categories is interesting because they can represent distinct cultural differences [15]. We show the total number of unique establishments collected in each city in Table 1.

Table 1. Number of unique establishments collected by city.

Cities	Curitiba	Rio de Janeiro	São Paulo	Chicago	New York	San Francisco
Yelp	1,755	2,046	2,964	3,652	4,280	3,340
Google	1,089	1,324	2,073	2,672	3,226	2,320

Analyzing Table 1 it is possible to notice that not all the establishments present in the Yelp list returned results of Google Popular Times during the search in Google. That is expected because Google does not offer this service for all venues. The HTML pages collected were processed to extract the popularity values for each day of the week. These values were then modeled as a discrete sequence of values v_k normalized between 0 and 1, where k represents each hour of the day so that a time series of popularity S can be defined as: $S = (v_k|\forall k \in [0\dots 23], 0 \leq v_k \leq 1)$, where for each establishment two time series are generated through the DBA technique, one representing its typical pattern on weekdays and other the pattern of weekends.

As the objective is to determine the most representative pattern exhibited by most of the time series in a cluster, anomalous members were removed by calculating the distance of each series to the centroid of the cluster and cutting the farthest apart. The cutoff threshold applied followed the rule $Quartile_3 + Interquartile * 1.5$, a conventional approach [5], from the centroid distribution, which resulted in the removal around 6%, on average, of the series in a cluster.

4 Clustering and Signatures Generation

4.1 Time Series Clustering

Identifying the behaviors, i.e., patterns of popularity, typically exhibited by each category of business may help answer whether the category has a unique homogeneous functioning dynamic or to verify if businesses present different hours of popularity peaks even belonging to the same category. For this purpose, companies of the same category with similar time series were detected and separated into clusters by applying the K-means algorithm with the DTW distance. The K-means algorithm has identified time series clusters more uniformly distributed than clusters generated through hierarchical partitioning. This approach has also been successfully explored in clustering time series of different domains [3,18].

Choosing the most appropriate number of clusters is a common challenge faced when performing unsupervised learning techniques on unclassified data [2]. The heuristic adopted in this study was the smallest number between the suggestion of the Silhouette method [11] and the Hartigan index [6]. This criterion is necessary due to the possibility that the two indices may not converge to the same value, and it can be justified by the little variation of signatures shapes resulted by increasing the number of clusters. Such a problem was found in [18] and [3], where the authors also used those two metrics.

We divided the time series clustering experiments into three stages: clustering of the time series of businesses in the same category (explained above), clustering of the signatures of different categories, and clustering of the categories signatures for different cities in a country.

The stage of clustering the signatures of different categories tries to identify behaviors displayed by more than one category, i.e., if it is possible to represent distinct categories with a single signature. For that, it is used the same criterion for choosing the number of clusters in the clustering of the time series of businesses in the same category.

Finally, a process to identify the signatures of countries, by associating the most similar categories of signatures among the cities, was performed. We consider the behavior of a category as representative of the country's behavior if, and only if, such behavior is present in all cities of the sample. For this reason, we determine the number of clusters of country signatures in each category by the number of signatures of the city with the least amount of clusters in that category.

4.2   Signature Generation

We can perform the generation of signatures to represent a set of time series in different ways depending on how the similarity between the time series is interpreted in the domain of interest. In the context of business popularity, we assume that time series are similar if they have similar shapes. Thus, a distance such as Euclidean would not be adequate for this scenario as the comparison occurs point by point; and two sets of time series with similar shapes of behavior separated by an offset of one hour might end up being separated when using this distance.

An alternative measure of distance that manages to overcome this problem is the Dynamic Time Warping (DTW). It tries to find the best global alignment between two time series such that, for example, a time series with a narrow peak that starts at 11h00 and ends at 13h00 is aligned with another time series containing a narrow peak that starts at 21h00 and ends at 23h00. To avoid the pitfall of the global alignment in the present scenario where time series with similar shapes in very distinct hours are aligned, a 2-h window has been applied to limit the alignment between two series.

The signature that represents a cluster of time series is generated using the DBA method which, in our case, starts with the centroid defined as the average observed movement by hour among all the time series of the cluster. Then it proceeds to a phase of iterative calculation of new centroid by aligning each time series of the set and the current centroid, so that at the end of execution the centroid represents an artificial time series that best aligns with the members of the set.

5   Results

5.1   Clusters and Signatures

The number of groups of businesses in a category suggested by the considered heuristic, that is, the number of existing distinct popularity patterns in the same category was usually between 2 and 4 for both weekdays and weekdays. To verify the actual need for more than one group per category a unique signature by category was generated initially.

In general, by generating only one signature representative of all the time series of businesses in a category, that is, without separating them into clusters, similar patterns of behavior still emerge when examining cities within the same country. We show representative signatures of each category on weekdays and weekends in Figs. 1 and 2 for Brazil and Figs. 3 and 4 for the United States, where the x-axis is the hour of the day, and the y-axis is the popularity. When comparing these results, it is possible to note that one category with great distinction is dance club, which we could expect because those type of places tends to open mainly at night, including on weekends.

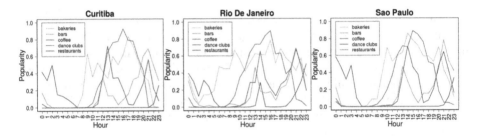

Fig. 1. Signatures for Brazil - weekdays.

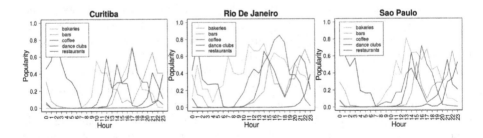

Fig. 2. Signatures for Brazil - weekends.

The drawback of using only one signature per category is the loss of the possibility of identifying distinct behaviors exhibited by each category. We illustrate that for the bar category in São Paulo. Figures 5 and 6 display several

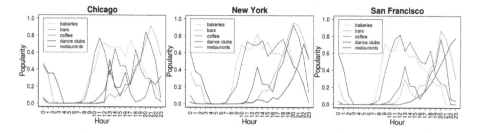

Fig. 3. Signatures for the United States - weekdays.

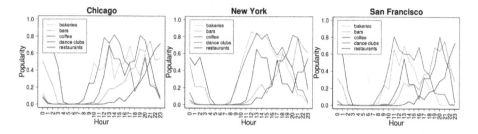

Fig. 4. Signatures for the United States - weekends.

time series of businesses in color and the generated signature in dotted black line for the two distinct groups discovered through the *K-means* algorithm for the category bar in the city of São Paulo on weekdays. It is noticeable that bars indeed have two distinctive types of behavior, the first presents more expressive peaks at 12h00 and 21h00 (9 p.m.) and the second with only one peak at 21h00 that decreases through the rest of the night. When studying those signatures for bar in that city, the signature containing only one expressive peak of visitation was not particularly visible in Fig. 1 (without clustering).

Analyzing the different signatures presented by a category, we can note that some are similar even though they belong to different categories. That can indicate the existence of businesses which, although officially declared as a specific category, have characteristics that are closer to another category. To investigate this phenomenon we clustered the category signatures to find out if signatures of the same category would form the clusters or resulted from the merging of different categories. As shown in Figs. 7 and 8, for example, the categories signatures of bar and dance club were grouped in the same set for the cities of Curitiba and Chicago.

This information could be useful in several tasks, for example, in the detection of establishments mistakenly registered in an inappropriate category. Also in the conception of meta-categories, or mixed categories, that more accurately describe the real behavior of a type of business, as the meta-category bar-dance club for establishments that although are initially bars are recognized as dance clubs by the visitors.

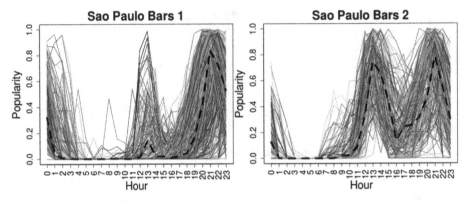

Fig. 5. Bars 1 - weekdays.   **Fig. 6.** Bars 2 - weekdays.

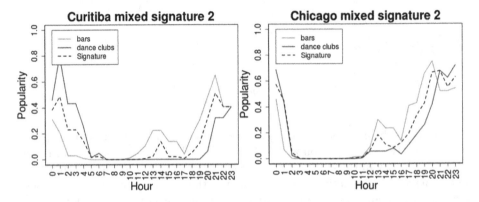

Fig. 7. Signature bar-dance clubs in Curitiba.   **Fig. 8.** Signature bar-dance clubs in Chicago.

To confirm that the behaviors found in each city are consistent within a country, that is, if there are patterns of popularity that typically occurs in a country's urban scenarios, the category signatures of each city were associated by minimizing the DTW distance among them. Studying these signatures generated, we can see that there are behavioral patterns that almost not differ among different urban scenarios of a country, independently of the city.

While the signatures representing bakery and coffee shop, Figs. 9 and 10, respectively, denote similar behaviors between Brazil and the United States, the signatures for bar and dance clubs, Figs. 11 and 12, respectively, display more differences between the two countries. Brazil exhibits a bar signature with the highest peak during the afternoon, and the United States presents a dance club signature with a peak earlier in the evening.

The restaurant category revealed two very distinct behaviors between the two countries. According to Figs. 13 and 14, while Brazilians tend to frequent restaurants mostly during lunch time, Americans prefer to visit this type of

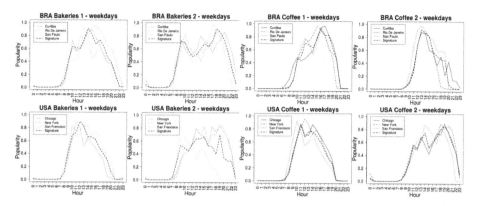

Fig. 9. Bakery signatures - weekdays.   **Fig. 10.** Coffee signatures - weekdays.

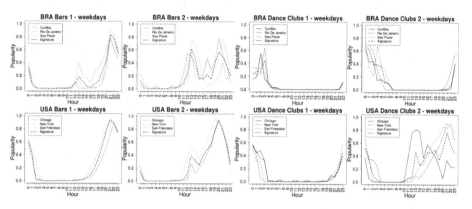

Fig. 11. Bar signatures - weekdays.   **Fig. 12.** Dance club signatures - week-days.

business at dinner. The same information was also observed using Foursquare check-ins [15].

These similarities and differences have important implications. The way in which people from different countries use certain businesses categories can be affected by many economic and social factors. Therefore the distance between country signatures could be explored when studying the culture of each country by using it as an index of similarity between countries.

5.2   Category Inference

In this section, we present the task of category inference of businesses using its time series and location as an example of application. To evaluate how much the signatures can generalize the behaviors of the categories, we compared two classification models.

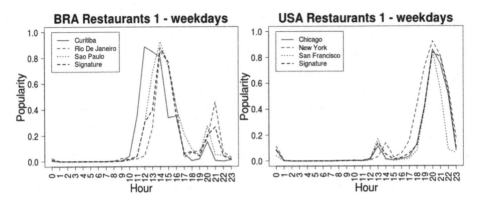

Fig. 13. Signature of restaurant in Brazil - weekdays.

Fig. 14. Signature of restaurant in the United States - weekdays.

The model $M1$ is based on a decision tree classifier that infers the category of an establishment given as attributes its time series, whether it represents the weekday or weekend, and the city from which it was obtained. We used the 10-fold cross-validation method to check the performance of the classifier in the set of all 12,704 time series. A second inference model, $M2$, was developed using the 87 signatures of categories and meta-categories, where an establishment is categorized according to the signature closest to its time series using the DTW distance, but with a small tolerance that favors the classification as a meta-category.

Table 2 presents the results for the performance comparison between $M1$ and $M2$ models. According to those values, it is noticeable that model $M1$ presented better performance in the F1-score, while model $M2$ obtained better precision on weekdays and accuracy on weekends. We might explain this result by the small loss of generality produced by the process of assigning a signature to a large set of time series. It is also possible to note that the performance in both models is slightly worse for weekends, which may indicate the existence of a lower consistency in the behavior performed by visitors compared to working days.

Table 2. Performance comparison between M1 and M2 models.

	Weekdays		Weekends	
	M1	M2	M1	M2
Accuracy	0.70	0.67	0.59	0.65
F1-score	0.69	0.63	0.66	0.60
Precision	0.64	0.66	0.67	0.62
Recall	0.76	0.60	0.66	0.58

As the performance presented by both models are similar, the use of signatures allows us to represent several different popularity times of commercial establishments using a reduced number of time series with little loss of information.

6 Meta-category Validation

One question that naturally emerges is whether meta-categories are useful to represent the perception of users regarding certain businesses. For that, an experiment was conducted to study the meta-category signatures with the help of ten volunteers aged between 18 and 30 years. They were instructed to respond to a multiple choice online questionnaire in which they should select categories from a group of five that could best describe the business. We also provided a free text option so that users were able to point out any supplementary information they thought necessary. Volunteers were free to interpret what defines each category. For each country, we randomly selected ten businesses that had been classified by a meta-category during the inference task, and, for each one, links for Facebook, Foursquare and Google were pointed to the volunteers so they could perform brief research about the venue to help them classify the venues.

Considering only the answers with the highest number of votes for the 20 analyzed businesses, 17 corresponded to the category considered as our ground truth, that is, the businesses category on the Yelp platform, while for the three remaining businesses the right category received the second highest number of votes. Therefore, according to the sample, the main categorization corresponds to what we would expect.

However, when analyzing all the response options that more than three people agreed on, we noticed that 14 businesses were classified under more than one category, corresponding to our meta-category classification. Therefore, the main category is not able to completely describe the business as the users perceive them. We can observe that in Table 3 which shows the evaluation of the volunteers about a bakery business that ended up receiving six votes for the coffee category.

Table 3. Classification of a bakery by volunteers

Category	Bakery	Bar	Coffee	Dance club	Restaurant
Votes	9	1	6	0	1

The six cases in which categories attributed by volunteers did not entirely match the meta-category appear to be due to confusion in the interpretation of the café category, in which the free-text option included comments such as "tea house", "ice cream shop", as well as questions about the definitions of the café category. This phenomenon sometimes also occurs in the classification used in websites, for example, we found cases where Google classifies certain businesses as a particular category, and the same place is described as a different category in Foursquare.

During the validation step, we discovered that the semantic of each category is difficult to define. Also, as we can note by the signatures of each country shown in Sect. 4, the interpretation of the meaning of each category seems to be affected by the cultural context of each country, so that only translating category names might not be sufficient to transmit the desired meaning completely. Therefore, the use of approaches such as the one presented in this study can be used to enrich the description of businesses by including knowledge about the behavior of their users.

7   Conclusion

In this study, information of the distribution of visits to commercial establishments in different cities in the United States and Brazil was collected from the Google Popular Times and used to discover patterns of popularity (signatures) that represent a group of establishments. When we analyzed the categories signatures, we could observe an association between different categories that we could use for the creation of meta-categories, perhaps more informative about the venue. Through an experiment with volunteers, we obtained evidence that those meta-categories make sense based on the perception of users. As one of the possible applications, we show that it is possible to infer the category of an establishment from its time series and city of location. For this purpose, we compared two inference models: one using the all the time series of commercial establishments and the second only the signatures of categories and meta-categories. Although using only the signatures did not have significant performance gains, using them reduces the required number of time series to perform this task. Besides, having signatures representing popularity times in different geographical regions, such as countries or cities, could enable cross-cultural studies regarding the visits time people tend to frequent business.

As future work, we propose the application of the signature discovery approach in a more substantial number of countries and cities, in order to identify possible similarities between them and explore the insights, for instance, in the better business description in different places. Another opportunity is to develop new services, for example, one that focuses on identifying and predicting low-movement hours, which can be exploited to facilitate the access to several services in fields such as banking, healthcare, and shopping.

Acknowledgements. This work was partially supported by the project CNPq-URBCOMP (process 403260/2016-7), CAPES, and Fundação Araucária. The authors would also like to thank all the volunteers for the valuable help in this study.

References

1. Arbelaitz, O., Gurrutxaga, I., Muguerza, J., PéRez, J.M., Perona, I.: An extensive comparative study of cluster validity indices. Pattern Recogn. **46**(1), 243–256 (2013)
2. Davies, D.L., Bouldin, D.W.: A cluster separation measure. IEEE Trans. Pattern Anal. Mach. Intell. **2**, 224–227 (1979)
3. Figueiredo, F., Almeida, J.M., Gonçalves, M.A., Benevenuto, F.: On the dynamics of social media popularity: a Youtube case study. ACM Trans. Internet Technol. (TOIT) **14**(4), 24 (2014)
4. Google: Google popular times (2017). https://support.google.com/business/answer/6263531. Accessed 10 Sept 2017
5. Han, J., Pei, J., Kamber, M.: Data Mining: Concepts and Techniques. Elsevier, Amsterdam (2011)
6. Hartigan, J.A.: Clustering Algorithms, vol. 209. Wiley, New York (1975)
7. Krumm, J., Rouhana, D.: Placer: semantic place labels from diary data. In: Proceedings of the 2013 ACM International Joint Conference on Pervasive and Ubiquitous Computing, Zurich, Switzerland, pp. 163–172. ACM (2013)
8. Liao, T.W.: Clustering of time series data - a survey. Pattern Recogn. **38**(11), 1857–1874 (2005)
9. Neves, Y.C., Sindeaux, M.P., Souza, W., Kozievitch, N.P., Loureiro, A.A., Silva, T.H.: Study of Google popularity times series for commercial establishments of Curitiba and Chicago. In: Proceedings of the 22nd Brazilian Symposium on Multimedia and the Web, Teresina, Piauí, Brazil, pp. 303–310. ACM (2016)
10. Petitjean, F., Ketterlin, A., Gançarski, P.: A global averaging method for dynamic time warping, with applications to clustering. Pattern Recogn. **44**(3), 678–693 (2011)
11. Rousseeuw, P.J.: Silhouettes: a graphical aid to the interpretation and validation of cluster analysis. J. Comput. Appl. Math. **20**, 53–65 (1987)
12. Salvador, S., Chan, P.: Toward accurate dynamic time warping in linear time and space. Intell. Data Anal. **11**(5), 561–580 (2007)
13. Silva, T.H., Loureiro, A.A.: Users in the urban sensing process: challenges and research opportunities. In: Pervasive Computing: Next Generation Platforms for Intelligent Data Collection, pp. 45–95. Academic Press (2016)
14. Silva, T.H., Vaz de Melo, P.O.S., Almeida, J.M., Salles, J., Loureiro, A.A.F.: A picture of Instagram is worth more than a thousand words: workload characterization and application, pp. 123–132, May 2013
15. Silva, T.H., de Melo, P.O.V., Almeida, J.M., Musolesi, M., Loureiro, A.A.: A large-scale study of cultural differences using urban data about eating and drinking preferences. Inf. Syst. **72**(Suppl. C), 95–116 (2017). https://doi.org/10.1016/j.is.2017.10.002, http://www.sciencedirect.com/science/article/pii/S0306437917300261
16. Tostes, A.I.J., Silva, T.H., Assuncao, R., Duarte-Figueiredo, F.L.P., Loureiro, A.A.F.: Strip: a short-term traffic jam prediction based on logistic regression. In: 2016 IEEE 84th Vehicular Technology Conference (VTC-Fall), Montreal, Canada (2016)
17. Vaca, C.K., Quercia, D., Bonchi, F., Fraternali, P.: Taxonomy-based discovery and annotation of functional areas in the city. In: Proceedings of ICWSM 2015, Oxford, UK (2015)
18. Yang, J., Leskovec, J.: Patterns of temporal variation in online media. In: Proceedings of the Fourth ACM International Conference on Web Search and Data Mining, Kowloon, Hong Kong, pp. 177–186. ACM. Kowloon (2011)

19. Ye, M., Shou, D., Lee, W.C., Yin, P., Janowicz, K.: On the semantic annotation of places in location-based social networks. In: Proceedings of the 17th ACM SIGKDD International Conference on Knowledge Discovery and Data Mining, pp. 520–528. ACM. San Diego (2011)
20. Yelp: Yelp developers (2017). https://www.yelp.com/developers/documentation/v3. Accessed 10 Sept 2017
21. Zheng, Y., Capra, L., Wolfson, O., Yang, H.: Urban computing: concepts, methodologies, and applications. ACM Trans. Intell. Syst. Technol. (TIST) 5(3), 38 (2014)

Contemporary Social Problems

Comparing Emotional Reactions to Terrorism Events on Twitter

Jonathas G. D. Harb and Karin Becker(⊠)

Instituto de Informática, Universidade Federal do Rio Grande do Sul (UFRGS),
Caixa Postal 15.064, Porto Alegre, RS 91.501-970, Brazil
jonathasgabriel05@gmail.com, karin.becker@inf.ufrgs.br
http://www.inf.ufrgs.br/~kbecker

Abstract. Over the last years, terrorism attempts have threatened the global population safety, impacting people in a complex emotional way. In this paper, we apply deep learning techniques to classify emotions of terrorism events, and develop a comparative analysis about emotional reactions on four events based on the demographics of tweeters, particularly gender, age and location. Our research questions involve comparing these events in terms of emotional shift, emotions according to age and gender, emotional reaction according to the closeness of the event and number/type of victims, as well as the terms used to express emotional reactions. The main conclusions were: fear, anger and sadness are the most expressed emotions; the emotions can be related to gender (e.g. fear for women, and anger for men); emotions seem to be not related to the closeness of the events, but seem to be affected by the casualties (number of kills/injuries); tweeters expressing fear and sadness tend to share words of affection and support, while tweeters expressing anger tend to use intense words of hate, intolerance and anger.

Keywords: Emotion analysis · Twitter · Deep learning

1 Introduction

Social media platforms such as Twitter are receiving increasing attention from researchers, as huge volumes of valuable information can be extracted at low latency and cost. Twitter allows millions of users to interact and express thoughts and opinions on a wide range of topics, producing large datasets of tweets and related user information, which can be exploited for several purposes. Twitter is becoming the first medium where people connect to express their emotions about important events, including terrorism attempts [2,23].

Over the last years, the variety of terrorism events has threaten the global population safety. These events are likely to impact people in a complex emotional way[1] as their goal is to outreach as many people as possible in a single strike, thus propagating constant fear and insecurity to a larger portion of the

---

[1] https://www.paulekman.com/blog/our-emotional-reactions-terrorism/.

© Springer Nature Switzerland AG 2019
J. Oliveira et al. (Eds.): BiDU 2018, CCIS 926, pp. 107–122, 2019.
https://doi.org/10.1007/978-3-030-11238-7_7

population [10]. In this context, exploiting the huge amount of data produced on Twitter to study the people's complex emotional reaction presents itself as an interesting opportunity in the field of sentiment analysis. Studies on emotional reaction to terrorism events can be helpful for developing assistance programs that provide support and help to better cope with terror [8].

Sentiment Analysis studies computational techniques to extract people's opinions, attitude and emotions from written language such as in tweets [13]. Sentiment analysis has been explored on terrorism-related tweets, but restricted to polarity analysis, i.e. sentiment classified as positive and negative. Some studies focused on modeling the information diffusion on Twitter about terrorism events [2,8,23], concluding that sentiment is a key factor on the size and survival of information flows. Emotion mining expands the sentiment analysis field by addressing the identification in texts of other affect states, as defined by emotion models [20] (e.g. basic emotions [4], such as anger or fear). Sentiments have been examined on Tweeter for understanding different social phenomena (e.g. [3,5,12]), but terrorism specific studies are still lacking.

Emotion analysis on Twitter is difficult due to the limited size of the texts, written in informal, internet-specific language, possibly with many errors. Common approaches for sentiment classification are the use of sentiment lexicons and machine learning methods [13]. The former is dependent on the existence of a dictionary that properly associates terms to sentiments in the domain. For emotions, a popular generic dictionary is the general-purpose NRC [17], which does not take into account social media language. The results of machine learning approaches are dependent on the existence of a domain-oriented annotated corpus of significant size as training set. Deep learning is a machine learning approach that involves automatic learning procedures that make use of computing power and large sets of data, minimizing the absence of large training sets [1].

Our research focuses on studying the emotional reaction of Twitter users to terrorism events. In this paper, we apply deep learning techniques to classify emotions of four (4) terrorism events, and develop a comparative analysis about emotional reactions based on the demographics of the tweeters, particularly gender, age and location. The emotion classification model was developed by training a Convolutional Neural Network (CNN) [11] over a terrorism-related tweet dataset. We expand our previous work [9] by considering terrorism events that occurred in two distinct countries, the United Kingdom (UK) and the United States of America (US), and by considering the following research questions:

- Q1: Is there an emotion shift due to terrorism events?
- Q2: Do different terrorism events raise the same emotional reaction?
- Q3: Does the user closeness to the event have an impact on the emotional reaction?
- Q4: Does the number and type of casualties have an impact on the emotional reaction?
- Q5: Are there differences on emotions' expression according to the events?

We analyzed two events in UK, and two in US, listed in Table 1. In the Manch-ester Arena bombing[2], a bomb was detonated by the end of Ariana Grande's concert. In the London Bridge attack[3], a van hit passing by pedestrians. In the New York City truck attack[4], a truck hit cyclists and runners. In the New York City attempted bombing[5], a bomb was partially detonated in a subway station. These events had distinct targets and casualties, and were widely publicized in the media. Thus, they are suitable for our comparative study.

Table 1. Overall information on the terrorism events

Event name	Location	Date	Victims
#prayformanchester	Manchester-UK	22 May 2017	22 kills/60 injuried
#londonbridge	London-UK	3 June 2017	8 kills/48 injuried
NY-october	New York-US	31 October 2017	8 kills/11 injuried
NY-december	New York-US	11 December 2017	0 kills/4 injuried

The main conclusions of our study are: there is indeed an emotion shift, and fear, anger and sadness are the most expressed emotions; emotions expressed are directly related to gender (fear for women, and anger for men); emotions seem not to be related to the closeness of the events, but seem to be affected by the casualties; tweeters expressing fear and sadness tend to regard themselves as potential victims, sharing words of affection and support, while tweeters express-ing anger tend to use intense words of hate, intolerance and anger.

The remaining of this paper is structured as follows. Section 2 describes related work. Section 3 describes the methods and materials used for providing answers to our research questions. Section 4 describes our experiments to find a suitable model for emotion prediction. Section 5 presents the analysis performed over the data to answer our questions. Finally Sect. 6 presents the conclusion and opportunities for future works.

2   Related Work

Sentiment analysis was target of several works for different purposes. A series of approaches for sentiment classification and data labelling were presented in [3, 5, 14, 16]. Kim in [11] applied deep learning in natural language text by training a Convolutional Neural Network (CNN) on top of pre-trained word embeddings. Models were evaluated against several datasets and the results outperformed several state of the art methods in the majority of the experiments.

---

[2] https://en.wikipedia.org/wiki/Manchester_Arena_bombing.
[3] https://en.wikipedia.org/wiki/June_2017_London_Bridge_attack.
[4] https://en.wikipedia.org/wiki/2017_New_York_City_truck_attack.
[5] https://en.wikipedia.org/wiki/2017_New_York_City_attempted_bombing.

Different works related demographics of tweets and sentiment to social phenomena. Lerman et al. in [12] related the structure of Twitter connections and the sentiment of tweets, and identified geo-located groups with predominant positive/negative emotions. In a related study [7], they analyzed the properties (such as expressed sentiments) of tweets in US metropolitan areas where people checked-in using Foursquare. Their results revealed that areas with many check-ins have happier tweets, encouraging other people to frequent these places.

Other works explored twitters' demographics to understand social phenomena. For instance, a study [5] analyzed engagement on the cause of Gender-based violence (GBV) based on demograhics and linguistic attributes, including emotions. Face++[6] was used to extract age and gender. They concluded that gender and age determines the engagement, and that anger is the predominant emotion.

Sentiment analysis has been explored in terrorism-related tweets, but restricted to polarity analysis. Studies focused on modeling the information diffusion [2,8,23] concluded that negative tweets largely outnumbers the positive ones, and that time lags between retweets, and the sentiment expressed are predictive of both size and survival of information flows. Regarding an UK event (Woolwich), the authors in [23] observed that negative and tense content is soon substituted by positive and supporting content. An experiment compared several sentiment classification approaches for ISIS-related tweets [15].

To the best of our knowledge, no work has addressed emotions in terrorism-related events, nor have analyzed the emotional reaction according to demographics and location. We deploy deep learning techniques as a means to obtain more accurate results and overcome the lack of large annotated datasets.

3 Materials and Methods

3.1 Dataset

We targeted four terrorism events that occurred in two different countries, summarized in Table 1. This choice was motivated by two factors. First, we focused on English written tweets in order to disregard differences due to the idiom, as well as to benefit from many tools and functions available for natural language processing. Second, to be able to understand the emotional reactions according to varied demographics and conditions, as the events happened in three different cities, varied in terms of casualties, and impacted people with different profiles.

The investigation of emotional reactions needs to contrast tweets prior to the terrorism event, and after it, such that we can analyze not only these reactions, but also a possible emotion shift due to the events. For each targeted event, we collected tweets two days before the event, the actual day it happened, and two days after the event. As the occurrence of a terrorism event is unpredictable, data collection must involve tweets from the past. We used an open source project[7] written in Python to overcome the restriction of the Twitter official streaming

---

[6] https://www.faceplusplus.com/.

[7] https://github.com/Jefferson-Henrique/GetOldTweets-python.

API, which does not allow to collect an unlimited number of tweets from the past. As parameters, we set query search terms combined with boundary dates. To define search terms to collect tweets, we analyzed raw data gathered from the web, trending topics, as well as samples extracted using the official Twitter API on the respective dates. We found recurrent hashtags for each one of the events, shown in Table 2, and used them as query search terms. Tweets containing these hashtags composed our terrorism event tweets (AFTER events). To compose our negative class (BEFORE events), we queried tweets using the generic keywords "Manchester", "London" and "New York", within the period of two days that preceded each event. We observed that these keywords were commonly used to tweet about citizen's thoughts on diverse topics such as football teams, universities, and daily news regarding these cities.

Table 2. Query terms, dates and dataset per event

Event Name	Query terms	Period	BEFORE (#tweets)	AFTER (#tweets)
#prayformanchester	#prayformanchester, "manchester"	05-20-2017 to 05-24-2017	BM (5,351)	AM (25,010)
#londonbridge	#LondonBridge, "London"	06-01-2017 to 06-05-2017	BL (20,379)	AL (29,656)
NY-october	#NYCStrong, #manhattanattack, "new york"	10-29-2017 to 11-02-2017	BNYO (12,130)	ANYO (11,072)
NY-december	#nycexplosion, #nycbombing, "new york"	12-09-2017 to 12-13-2017	BNYD (64,906)	ANYD (3,602)

We performed typical data pre-processing actions, such as removal of hyperlinks, mentions, hashtags used to identify the events, special symbols (e.g. &, /), etc. We applied an English dictionary[8] to discard tweets with too many misspelling, as well as non English ones. We produced eight datasets[9]: BM (before Manchester), BL (before London), BNYO (before New York October), BNYD (before New York December), AM (after Manchester), AL (after London), ANYO (after New York October) and ANYD (after New York December). The number of tweets for each dataset is shown in Table 2. The structures of these datasets are identical and include the tweet ID, text and metadata.

Afterwards, we related each tweet to the respective demographics in terms of location, gender and age of the respective tweeter. Given that less than 1% of the collected tweets were geo-referenced, we assumed location as informed in the user's profile, a commonly deployed technique to overcome this issue [22]. However, the location in each user profile is a free text informed by the user without any validation. Thus, we compared each declared location against a list of cities from UK and US. To identify age and gender, we used Face++ as in [5], based on the experiments that report an accuracy of 85% [6].

Our original plan was to analyze the sentiment per city, but the number of tweets for comparison, given such a location granularity, was very small. Alternatively, a representative number of tweets could be taken into account if we grouped different locations in a higher level abstraction. Therefore, we generalized the original locations into the following categories: (a) for the UK events, we

---

[8] https://github.com/dwyl/english-words.
[9] Public available at https://github.com/jonathasgabriel/Terrorism-Twitter-Dataset.

considered locations from the UK, locations from the US, and "other locations"; (b) for the US events, we considered locations from New York, locations from the US other than New York, and "other locations".

3.2 Emotion Gold Standard

Our work focuses on five out of the six basic emotion categories defined by Eckman [4]. The emotion categories considered include anger, fear, sadness, surprise and disgust. We focused on negative emotions only, because we assumed that people are not likely to express happiness in reaction to terrorism events (See footnote 1). Our approach considers that a given tweet is related to one and only one of the emotion categories, regarded as the predominant emotion.

Table 3. Gold standard: number of labelled tweets per category

Emotion	Anger	Disgust	Fear	Sadness	Surprise	None
# tweets	82	116	85	179	71	74

To train a model for emotion prediction, an emotion labeled dataset is required. As domain-related datasets tend to provide the best results [13], we created a specific terrorism gold standard for the task. Tweets were labeled according to each emotion category considered, plus an extra "none" category. This was accomplished using Amazon Mechanical Turk[10].

First, one of the authors annotated 967 tweets that were likely to be in each of the categories due to the presence of emotion keywords and expressions. In the annotation process, the author distributed the tweets as evenly as possible into the categories. Emotion keywords were defined by data sampling. For example, the tweet *"Deeply saddened by the loss of 22 beautiful lives. We should not live like this."* was labelled as sadness due to the expression "deeply saddened". On the other hand, the tweet *"It's so scary to not feel safe in this World"* was labelled as fear due to the expression "It's so scary", and so on. Afterwards, we created a HIT (Human Intelligence Task) with these tweets, where annotators were asked to determine which emotion best described a tweet, given a set of categories as options (anger, fear, disgust, sadness, surprise and none). We instructed annotators to choose the primary emotion if more than one emotion could be identified, and to choose "none" if no emotion could be clearly determined. We targeted the HIT to two master annotators, so that we would have three annotators in total, including the annotator author. We filtered out tweets in which there was a disagreement between all the three annotators, and retained those with at least two agreements, used as labels. The results, composed of 607 tweets, are displayed in Table 3, which we consider as our ground truth for validating the emotion prediction model.

---

[10] https://www.mturk.com/.

3.3   Classification

In order to classify our collection of tweets, we applied deep learning by training a Convolutional Neural Network (CNN) as defined in [11]. Our choice is due to their results, and the pioneering in using such approach for classifying natural language. Another motivating factor for using such an approach was the automatic learning capability that deep learning has by incorporating improved learning procedures that make use of computing power and training data, working well on large sets of data [1].

In a nutshell, the CNN architecture comprises four layers. The first layer converts words into vectors of low-dimensional representation called *word embeddings*. The second layer applies a series of convolutions over these word embeddings to produce a feature map for each sentence. The third layer is responsible for filtering the most important features into one feature vector through a max polling operation. The fourth layer applies the softmax function to classify sentences into labels. The Python code of the CNN implementation we used is publicly available[11,12], and it is designed to be executed on the top of TensorFlow[13]. Our initial experiences with this CNN for emotion classification are described in our previous work [9].

4   Emotion Classification Experiments

We developed experiments with our CNN to find the most suitable classification model for our emotion categories. The CNN parameters we used were the same as in [11] because their results were built using these parameters. In fact, we tried variations on these parameters, but no significant differences were observed.

Given that our limited number of labelled tweets did not provide enough data for properly training the CNN, we tried different approaches for gathering training seeds. We experimented two strategies, namely *distant supervision* [24], using an emotion-labelled dataset containing electoral tweets [19] and *filtering* tweets from the dataset with specific properties and use them as training seeds. We experimented three filtering criteria: (a) *hashtags* containing emotion names (e.g. angry) [17]; (b) *emotion words* contained in the sentiment lexicon NRC [18]; and c) frequent and co-occurring *keywords* extracted from samples of our dataset. For each strategy, the CNN was trained and a prediction model was generated. The test was always conducted against our labelled Gold Standard. From all of our experiments, the one using filtering by keywords provided the best results, and was adopted to generate the emotion prediction model. Further details on these comparisons are provided in [9].

---

[11] https://github.com/cahya-wirawan/cnn-text-classification-tf.
[12] https://github.com/dennybritz/cnn-text-classification-tf.
[13] https://www.tensorflow.org/.

The keywords used to filter out the training seeds are summarized in Table 4. To improve our results, we loaded in our CNN pre-trained word embeddings, a mean of improving performance when the training set is not large enough, as suggested in [11]. We chose the word embeddings corpus provided by GloVe[14] because it is extracted specifically from tweets. Incorporating the GloVe's embedding set in the CNN improved our results.

Table 4. Keywords used for filtering training seeds for the CNN

Emotion	Keywords
Anger	Anger, fuck, fucked, pissed, lmaof, damm
Disgust	Disgust, disgusted, disgusting
Fear	Worried, worry, scary, scaring, scared, fear
Sadness	Sad, sadness, saddened
Surprise	Surprised, surprising, surprise, shocked, shocking

Table 5 displays our results per emotion, as well as macro-average measures. We considered our model reliable because it achieved average precision and recall above 70%, which we believe were good results taking into account results presented in related work on emotion classification [19,21,24].

Table 5. Results for the model generated by filtering keywords

Emotion	Anger	Disgust	Fear	Sadness	Surprise	None	Macro-avg
F-measure	0.86	0.65	0.64	0.73	0.61	0.66	0.65
Precision	0.80	1.00	0.55	0.86	0.74	0.49	0.77
Recall	0.93	0.48	0.76	0.64	0.51	1.00	0.70

5   Analysis

5.1   Q1: Is There an Emotion Shift Due to Terrorism Events?

To answer this question, we firstly compared the distribution of tweets with and without emotions before and after each event. Figure 1 depicts this comparison, where Y axis represents the percentage of tweets with regard to the total number of tweets, per event. We observed that, for all events, the same pattern is present: over 8% of the tweets that preceded the events contained emotions, compared to 25% after them.

---

[14] https://nlp.stanford.edu/projects/glove/.

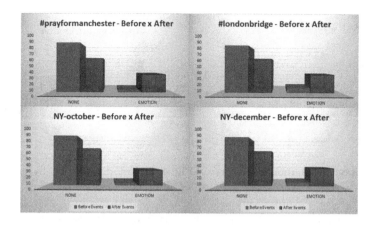

Fig. 1. Tweets distribution before events and after events.

Additionally, we compared the emotion distribution before and after each event. Figure 2 displays such comparison, where Y axis represents the percentage of the total number of tweets distributed into emotion categories, per event. We can observe that, after all events, notably three emotions prevailed: anger, fear and sadness. No significant changes were observed for disgust and surprise. Therefore, we conclude that there is indeed an emotional shift due to terrorism events, and that specifically anger, fear and sadness are the emotions that prevail once a terrorism event occurs.

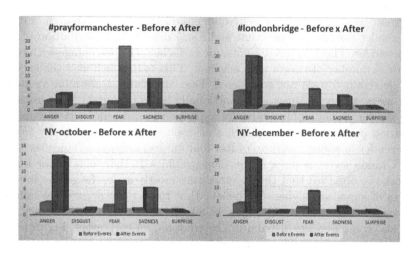

Fig. 2. Tweets distribution by emotion before and after events.

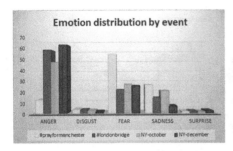

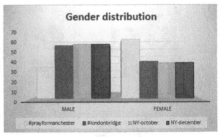

Fig. 3. Emotion distribution for terrorism events.

Fig. 4. Tweet distribution by gender for all events.

5.2 Q2: Do Different Terrorism Events Raise the Same Emotional Reaction?

To answer this question, we compared the distribution of emotions for each event (datasets AM, AL, ANYO, ANYD). Figure 3 depicts this comparison, where the Y axis represents, for each class (#prayformanchester, #londonbridge, NY-october and NY-december), the percentage of its total number of tweets distributed into emotion categories. Only tweets with emotion are displayed. These results reveal that there are differences in terms of emotion distribution per event. All events evoked mostly anger, except for the one in Manchester, where the most predominant emotion was fear. Notice that the New York December event (NY-december) evoked less sadness, compared to the other events. We believe that this is explained by the fact that this terrorism attack has not resulted in deaths.

We used demographics to help understanding these differences. Figure 4 details gender distribution by event and Fig. 5 depicts the distribution of emotions by gender. The Y axis represents, for each class, the percentage with regard to the total number of tweets with emotion. Tweets in which user gender could not be determined were not considered. Fear and sadness are more related to Female users, whereas Anger is proportionally more related to Male users. We believe gender demographics partially explains the differences on the emotion distribution per event. In the Manchester event, tweeters were mostly women, i.e. the gender in which measures of fear and sadness are higher, since the singer Ariana Grande is very popular in this demographics. In the events of London and New York, we observed that the tweeters were mostly male, the gender in which measures of anger are higher. We believe that these three events have affected the average citizen, who could be potentially at the location of the attack.

We also compared age distribution by emotion, for each event. Figure 6 illustrates such comparison. The Y axis represents, for each class, the percentage with regard to the total number of tweets with emotion. Tweets in which user age could not be determined were not considered in the comparison. It is possible to observe a pattern concerning the UK events. As the age increases, the feeling of anger increases proportionally. Fear, on the other hand, is higher for young ages, and it smoothly drops as age increases. Regarding the US events,

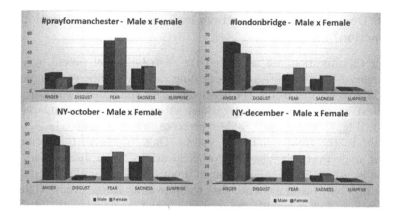

Fig. 5. Emotion distribution by gender for all events.

starting from age 19, there is a tendency of anger to remain constant while fear has the tendency to drop as age increases. Sadness showed no clear behavior in any event. Unlike for gender, we could not observe a pattern across the events, and thus the influence of age on emotional reaction was inconclusive.

With these analysis, our conclusion is that each terrorism event may raise distinct predominant emotions according to the demographics. However, in the analyzed events, gender was more influential on the emotion than age.

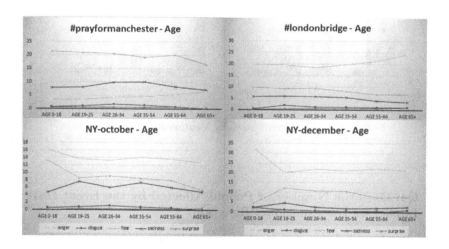

Fig. 6. Age distribution by emotion for all events.

5.3   Q3: Does User Location Have an Impact on the Emotional Reaction?

To answer this question, we compared the distribution of tweets by emotion for each defined location category. Regarding the UK events, we compared UK against US tweets. Regarding the US events, we compared tweets from New York against tweets from other locations within US. These comparisons are depicted in Fig. 7, where the Y axis represents, for each class, the percentage with regard to the total number of tweets. Only tweets with emotions were considered, and tweets classified as "other locations" were disregarded. For all considered locations, these comparisons did not reveal any noticeable difference. These findings may be an evidence that the proximity to the event, as indicated by the location, does not affect the evoked emotion as much as the gender who expresses it.

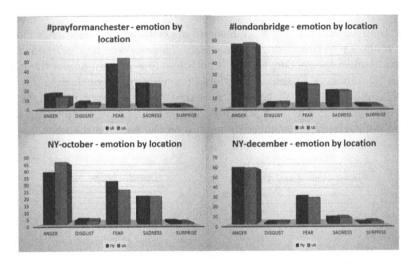

Fig. 7. Tweet distribution by location for the events.

5.4   Q4: Does the Number of People Affected by the Event Have an Impact on the Emotional Reaction?

To answer this question, we analyzed whether the number and type of victims affect the type of emotions expressed. Figure 8 displays the amount of anger, sadness and fear per event, since these are the three predominant emotions. Y axis represents, for each class (22k /60i, 8k/48i, 8k/11i and 0k/4i), the percentage of its total number of tweets distributed into emotions (anger, fear and sadness). Each class of kills/injuries represents the number and type of victims of each event, as shown in Table 1. We can observe that the level of sadness is directly related to the number of victims, whereas anger is inversely related. Fear is highly present when the number of victims is really high, but remains constant otherwise. Our hypothesis, to be confirmed, is that sadness might be a feeling more related to an event that victimizes more people, possibly because people feel helpless, or due to a sense of solidarity towards their families and close ones. Conversely, anger might be more present when tweeting about events that have fewer victims, due to the potential of harm. It is possible that these emotions are biased by the user's gender, and the conclusions with regard to fear may be more related to the demographics for the Manchester event than to the number and type of casualties. Another possibility with regard to fear is that, like sadness, people might relate to the victims and their loved ones, sharing a fear that they could be the victims themselves.

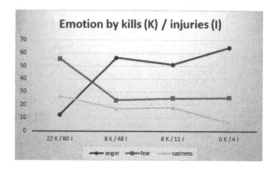

Fig. 8. Anger, fear and sadness behavior by number of victims.

5.5 Q5: Are There Differences on How Emotions are Expressed According to the Event?

Finally, we searched for common words/expressions used by tweeters to demonstrate emotions regarding the events. We focused on the three prevailing emotions (anger, fear and sadness) and created word clouds for each event, depicted in Fig. 9. We found patterns related to the events themselves (e.g. location and nature of the event), emotions and terrorism events in general.

The first noticed characteristic is that many of the keywords defined for filtering training seeds in our model (Table 4) were present in the respective emotion word clouds (e.g. "fuck" for anger, "worried" and "scared" for fear, "sad" and "saddened" for sadness). This is an evidence that the defined keyword are indeed representative.

We found common terms across all clouds that relate to terrorism events in general (e.g. "attack") and words related to religion (e.g. "muslim", "muslims", "islamic", "islam"), regardless of the emotion. Indeed, many people believe that Muslim extremists are behind these attacks, either as declared practitioners or as recruited members of an extremist religious terrorist group. Representative of each event are neutral terms describing the location, such as "manchester" and

Fig. 9. Word clouds for all events, by emotion.

"concert", "london" and "bridge", "new" and "york". The nature of the attack is also described accordingly, using "bomb", "bomber", "explosion", and "van".

With regard to the clouds for fear and sadness, we observe more frequent words that demonstrate solidarity such as "pray", "prayers" and "condolences". We suppose that tweeters expressing fear and sadness are likely to relate to the victims and their close ones, putting themselves in their shoes and sharing words of affection and support. On the other hand, clouds for anger show with frequency words of hate and intolerance, such as "stop", "terror", and "awful".

For the US events, the word "trump" (referencing the US president) was present in 4 out of the 6 word clouds and "obama" (former US president) appeared once. Such a political mention appear just once ("political") for UK events. We suppose that this may show a cultural difference between US and UK tweeters. US ones might be more engaged in asking politicians (or their President) for safety (or blaming them).

6   Conclusion

Our work provided a study on the emotional behavior of Twitter users in reaction to terrorism events. We addressed negative emotions and used deep learning approach for emotion prediction. Demographic data such as location, age and gender were extracted with help of available tools and helped explain our results. In answering our proposed questions, we showed that when terrorism events occur, a shift of emotion towards anger, sadness and fear can be noticed. In addition, our demographic analysis showed that gender have influence on tweeters reactions. Our data indicated that measures for fear and sadness are proportionally higher for Women than for Men, whereas anger is proportionally more related to the Male demographics. In addition, the number of affected people in each event might also have an impact. The feeling of sadness is directly proportional to the number of victims, where the feeling of anger is inversely proportional. Fear is predominant if the number of victims is really high, and tends to remain constant otherwise. The word clouds we created showed that words identifying event location, nature of the attack and emotions were commonly used. Emotions such as fear and sadness evoke the use of words that demonstrate solidarity, and emotions such as anger evoke the use of words of hate, intolerance and anger. Lastly, location analysis might indicate that it does not influence the evoked emotion as much as other factors, such as gender.

The terrorism tweet dataset we created is one of our contributions along with the techniques and tools we presented to gather our data and achieve our results. The questions we answered were a first step towards understanding the emotional reaction terrorism events raise on general population. The reported results might be used in developing specific assistance programs for coping with terror. We hope our work encourage further studies on social media focusing on terrorism, which we believe impact people in a complex emotional way. Finally, the data we derive might be used for further analysis we aim to do. One opportunity is to extend our approach to general attack events such as mass shootings that are becoming recurrent.

References

1. Ain, Q.T., et al.: Sentiment analysis using deep learning techniques: a review. Int. J. Adv. Comput. Sci. Appl. **8**(6), 424 (2017)
2. Burnap, P., Williams, M.L., Sloan, L., et al.: Tweeting the terror: modelling the social media reaction to the Woolwich terrorist attack. Soc. Netw. Anal. Min. **4**(1), 206 (2014)
3. De Choudhury, M., Jhaver, S., Sugar, B., Weber, I.: Social media participation in an activist movement for racial equality. In: Proceedings of the ICWSM, pp. 92–101 (2016)
4. Ekman, P., Friesen, W.: Emotion in the Human Face System. Cambridge University Press, San Francisco (1982)
5. ElSherief, M., Belding, E.M., Nguyen, D.: # notokay: Understanding gender-based violence in social media. In: Proceedings of the ICWSM, pp. 52–61 (2017)

6. Fan, H., Cao, Z., Jiang, Y., Yin, Q., Doudou, C.: Learning deep face representation. CoRR (2014). http://arxiv.org/abs/1403.2802

7. Gallegos, L., Lerman, K., Huang, A., Garcia, D.: Geography of emotion: where in a city are people happier? In: Proceedings of the WWW, pp. 569–574 (2016)

8. Garg, P., Garg, H., Ranga, V.: Sentiment analysis of the URI terror attack using Twitter. In: Proceedings of the International Conference on Computing, Communication and Automation (ICCCA), pp. 17–20 (2017)

9. Harb, J.G.D., Becker, K.: Emotion analysis of reaction to terrorism on Twitter. In: Proceedings of the SBC Brazilian Symposium on Databases, pp. 97–108 (2018)

10. Horgan, J.: The Psychology of Terrorism, 2nd edn. Taylor & Francis Group, Abingdon (2014)

11. Kim, Y.: Convolutional neural networks for sentence classification. In: Proceedings of EMNLP, pp. 1746–1751 (2014)

12. Lerman, K., Arora, M., Gallegos, L., Kumaraguru, P., Garcia, D.: Emotions, demographics and sociability in Twitter interactions. In: Proceedings of the ICWSM, pp. 201–210 (2016)

13. Liu, B.: Sentiment analysis and opinion mining. Synth. Lect. Hum. Lang. Technol. **5**(1), 1–167 (2012)

14. Lotan, G., et al.: The Arab Spring| the revolutions were tweeted: information flows during the 2011 Tunisian and Egyptian revolutions. Int. J. Commun. [S.l.] **5**, 31 (2011). https://ijoc.org/index.php/ijoc/article/view/1246. Accessed 26 Dec 2018. ISSN 1932-8036

15. Mirani, T.B., Sasi, S.: Sentiment analysis of ISIS related tweets using absolute location. In: 2016 International Conference on Computational Science and Computational Intelligence (CSCI), pp. 1140–1145 (2016)

16. Mitchell, L., Frank, M.R., Harris, K.D., Dodds, P.S., Danforth, C.M.: The geography of happiness: connecting Twitter sentiment and expression, demographics, and objective characteristics of place. PLOS ONE **8**(5), 1–15 (2013). https://doi.org/10.1371/journal.pone.0064417

17. Mohammad, S.: #Emotional tweets. In: Proceedings of the First Conference on Lexical and Computational Semantics, pp. 246–255 (2012)

18. Mohammad, S.M., Turney, P.D.: Crowdsourcing a word-emotion association lexicon. Comput. Intell. **29**(3), 436–465 (2013)

19. Mohammad, S.M., Zhu, X., Kiritchenko, S., Martin, J.: Sentiment, emotion, purpose, and style in electoral tweets. Inf. Process. **51**(4), 480–499 (2015)

20. Munezero, M.D., Montero, C.S., Sutinen, E., Pajunen, J.: Are they different? Affect, feeling, emotion, sentiment, and opinion detection in text. IEEE Trans. Affect. Comput. **5**(2), 101–111 (2014)

21. Purver, M., Battersby, S.: Experimenting with distant supervision for emotion classification. In: Proceedings of the 13th Conference of the European Chapter of the Association for Computational Linguistics, pp. 482–491 (2012)

22. Sakaki, T., Okazaki, M., Matsuo, Y.: Earthquake shakes twitter users: real-time event detection by social sensors. In: Proceedings of the 19th International Conference on World Wide Web, pp. 851–860 (2010)

23. Simon, T., Goldberg, A., Aharonson-Daniel, L., Leykin, D., Adini, B.: Twitter in the cross fire—the use of social media in the Westgate mall terror attack in Kenya. PloS ONE **9**, e104136 (2014)

24. Suttles, J., Ide, N.: Distant supervision for emotion classification with discrete binary values. In: Gelbukh, A. (ed.) CICLing 2013. LNCS, vol. 7817, pp. 121–136. Springer, Heidelberg (2013). https://doi.org/10.1007/978-3-642-37256-8_11

Using Government Data to Uncover Political Power and Influence of Contemporary Slavery Agents in Brazil

Letícia Dias Verona[(✉)] [iD], Giseli Rabello Lopes [iD],
and Maria Luiza Machado Campos [iD]

Programa de Pós-Graduação em Informática (PPGI),
Universidade Federal do Rio de Janeiro - (UFRJ), Rio de Janeiro, Brazil
leticiaverona@ufrj.br, giseli@dcc.ufrj.br,
mluiza@ppgi.ufrj.br

Abstract. This work uses open data published by the Brazilian government to investigate connections between agents involved on contemporary slavery labor and politicians, evaluating their power and influence. A network was built on data from Brazilian elections and campaign donations since 2002, including all candidates and donors associated to slave labor. Not only 263 direct candidatures from slavery agents were identified, but also more than 40 million Brazilian Reais in campaign donations for candidates for all electoral positions, showing a strong relation between slavery agents and Brazilian politicians. Data were also analyzed using metrics based on sociologist Manuel Castells' Network Theory of Power that measure how much power and influence each donation is accounted for, in addition to its absolute amount. The resulting network was semantically enriched and modeled according to existing ontologies and published in RDF using Linked Open Data standards in a semantic knowledge graph, allowing information to be identified, disambiguated and interconnected by software agents in future research.

Keywords: Knowledge graphs · Government open data ·
Social Network Analysis · Power and influence analysis

1 Introduction

The identification of power groups that act within the political structure is a tool for citizens and also a challenging research topic. This study aims to collaborate on the fight against contemporary slavery in Brazil. We build a knowledge graph to represent a network encompassing employers caught using slave labor and politicians connected to them. The graph is semantically enriched to deploy available domain knowledge and was derived from publicly published data from governmental and non-governmental organizations. The connections were made using electoral campaign donations and corporation partnerships. We apply metrics to measure political power in the graph, using this Brazilian study case as a motivation and experimentation base. Our strategy differs from previous works as we use a combination of topological features and

© Springer Nature Switzerland AG 2019
J. Oliveira et al. (Eds.): BiDU 2018, CCIS 926, pp. 123–138, 2019.
https://doi.org/10.1007/978-3-030-11238-7_8

domain knowledge, grounded on sociology theory, in a mixed approach that leverages results in a real world situation.

The use of big data to fight contemporary slavery is explored in many quantitative and qualitative researches [1, 2]. Despite the international agreements to ban slavery in the world, the crime persists. In Brazil, more than forty thousand people were rescued from slavery in the last decade. We could ask "How come that in the twenty-one century we still have enslaved people?", but this would sound as the outrage speaking. Instead, we propose the following research question: *How can we measure power in a political network in order to understand the forces that sustain the use of slave labor in Brazil?*

The fight against slavery in Brazil has a powerful instrument: the "Dirty List of Slavery". This list is published every six months by the Brazilian Labor Ministry and contains the names of employers that, after all administrative defense, could not prove innocence against the charge of slave labor use and, therefore, should loose access to public funding and face criminal charges. The list created a judiciary battle between involved agents and anti-slavery activists. The first ones try to block the publication of the list by all means, and the second ones fight to keep the list fully functional. Every now and then the news highlight that politicians are involved in slavery chains in Brazil, but a deeper and systematic analysis of these connections is necessary. Connections that have a practical outcome: there is no public available dataset of all editions of the Dirty List of Slavery and a historic compilation and evolution analysis is hard to accomplish.

This study presents a dataset modeled as a knowledge graph containing all editions of the Dirty List of Slavery, derived from three data sources. This complete dataset had not yet been published, only singular editions in old news, frequently in formats computers cannot process, like images.

To investigate the hypothesis that this old and recurrent crime must be sustained by political power, we add to the graph all Brazilian politician related to them using electoral campaign donations and corporation partnerships as relationships. Finally, we apply metrics designed to evaluate political power to this knowledge graph. The results show deep connections of Brazilian politicians and slavery agents in federal, state and municipal levels. The concentration of slavery episodes in some cities are not a coincidence, since their mayors and councilmen are themselves involved in illegal situations (practicing slavery in their farms or industries) or have strong connections with the crime, receiving campaign donations from these employers.

Concerning the structure of this paper, in Sect. 2, we present some definitions of contemporary slavery and a succinct explanation of Brazilian laws, as well the results as a description of the available data about slavery and campaign donations. In Sect. 3, we discuss the metrics used and the rational that guided the construction of the network. We also review concepts from the Semantic Web used to build the knowledge graph. In Sect. 4, we describe the process of data gathering, cleaning and transformation as well as analyzing data and we show the results of the analysis of the Brazilian political scenario. Finally, in Sect. 5, we highlight the limitations of our work and point out potential future work.

2 Data About Brazilian Politicians and Slavery in Brazil

In 2009, Brazil was one of the founders of the *Open Government* movement and, together with seven other countries, signed the first international transparency agreement, headed by the Unites States and later endorsed by more than 50 countries, that compromised to publish their data to guarantee free access to government information [3]. Beyond publishing the data, to **opening** them means to reduce the barriers to their use and reuse. The same agreement states that open data should: (i) be complete, not subject to access privileges; (ii) be primary, in the sense of granularity; (iii) be accessible; (iv) be machine readable; (v) be time sensible, in the sense of to be published in time to be useful; (vi) have free access without approval needs; and (vii) have an open license. Signatory of this agreement, the Brazilian government has published a huge amount of data available its government data portal [4]. It includes valuable datasets, but published in formats that are difficult to be processed and analyzed by computers, such as images and tables inside images. Moreover, most of them lack metadata to inform the real meaning of data item. Brazilian Transparency Law (Lei da Transparência, Law No. 12,527, 12/18/2011) also guarantees that all citizens have access to any public information unless explicitly classified as private, and public institutions must rapidly respond data requests.

2.1 Contemporary Slavery in Brazil and Available Data

The end of slavery and analogous forms of labor is a goal agreed by the international community. The International Labor Organization (ILO) has two conventions that address the issue: Forced Labor Convention, 1930 (No. 29) and The Abolition of Forced Labor Convention, 1957 (No. 105), both signed by Brazil. The first one states that all forced labor should be banished, except in particular cases: compulsory military service; normal civic obligations; prison labor (under certain conditions); work in emergency situations (such as war, calamity or threatened calamity) and minor communal services (within the community). The second one specifically prohibits the use of forced labor by State governments, especially as punishment for the expression of political views; for the purposes of economic development; as a means of labor discipline; as a punishment for participation in strikes and as a means of racial, religious or other discrimination. Both Forced Labor Conventions enjoy nearly universal ratification, meaning that almost all countries are legally obliged to respect their provisions and regularly report on them to the ILO's standards supervisory bodies. According to the ILO Conventions, forced or compulsory labor is: "all work or service which is exacted from any person under the threat of a penalty and for which the person has not offered himself or herself voluntarily" [5].

In Brazil, forced labor has been defined as a form of *modern-day slavery*. This includes debt bondage, degrading work conditions, and long work hours that pose a risk to workers health or life, and violate their dignity. The workers are usually taken to a geographically isolated place and cannot leave it, sometimes guarded by armed people.

But the strongest retention mechanism is illegal workers' debt. The public institution responsible for monitoring labor conditions is the Labor Department of Justice (Ministério Público do Trabalho). From 1994 to 2016 a number of government policies were created to fight contemporary slavery in Brazil. The **Dirty List of Slavery** is a conjunction of two administrative government ordinances: No. 504 from Ministry of Labor (Ministério do Trabalho e Emprego) and No. 1150, from Ministry of National Integration (Ministério da Integração Nacional) both issued in 2004. The first ordinance created the list and the second one recommends that financial agencies shall not grant public funding to agents in the list. The insertion in the list only happens when the administrative appeals are judged, assuring broad defense. Thus, within all constitutional precepts, the list, at the same time as it publishes information, removes from the agents of slave labor one of its greatest sources of wealth: obtaining rural credit from public banks [6]. The Dirty List of Slavery is the master data of this study, the main source in whose orbit other data will be worked. The list is composed of operations carried out by the Labor Department of Justice (Ministério Público do Trabalho) that caught slave labor practices. Each entry is therefore the description of an operation, and reveals the name of the employer or property owner responsible for the enslavement. An operation may appear on several editions of the list, since originally listed, and employers can only be removed from the list: (i) after complying with all public prosecution determinations, payment of indemnities and fines; (ii) if not reincident; and (iii) after a minimum period of two years. The Supreme Court (Supremo Tribunal Federal) suspended the list in 2014 and reissued it with revised removal criteria: an employer can sign a term of conduct adjustment and negotiate its removal from the list. The Dirty List of Slavery faces many opponents inside and outside the government. It should be published every six months, but judicial battles prevented this periodicity from being maintained. In addition, the data has not been available for a long time: although the Ministry of Labor publishes the list, there is no historical record of the previous lists. This work makes public a dataset with all editions of the list, using a standard format and adhering to all principles of the W3C Linked Open Data. Besides, it also finds and measures the connections between the enslavers in the list and public power in Brazil, using metrics designed for this purpose.

As there is no complete government data source, the historical basis was assembled using data from three sources. The first is the website of the Ministry of Labor. The survey has obtained all lists published directly on this site since 2014, immediately after their disclosure. Unfortunately, this action cannot be reproduced: invariably, a few days after publication, the list was not available. The second source is the anti-slavery NGO Repórter Brasil website [7] which, through injunctions based on the Transparency Law, has obtained editions of the list in the moments when the publication was suspended and the third source is data obtained directly from the religious organization Comissão Pastoral da Terra [8] (a strong force against slave labor and other forms of exploration in Brazil) which made available for the research a list compiled from its creation in 2003 until 2013. Transformation steps were necessary to use tools to convert data obtained from different file formats (e.g. PDF or images) to machine-readable data in text formats.

For the purpose of this research, as the interest is to assemble a history of the data, each edition of the list was identified, and therefore it is possible to see if the same employer is recurring in multiple editions of the list.

2.2 Data About Elections and Campaign Donations

Considering the Brazilian Election Scenario, candidates for public legislative or executive positions must deliver the report of their expenses as well as income during electoral campaigns. But these raw data are difficult to visualize and analyze. For example, data available in the Electoral Repository [9] have changed over time, containing more information for each year of election. For the year 1998, campaign donations are not published, only the registration of candidacies. In the elections from 2002 to 2008, the CPFs of the candidates[1] were not available, which makes it very difficult to cross information. This was circumvented by the fact that the vast majority of candidates continued in public life and it was possible to retrieve these CPFs by full name combined with the candidates' date of birth.

Donations from corporations, once legal, were banned in the 2016 elections and in this election only donations of individuals up to the limit of ten thousand Brazilian Reais were accepted. In this study, we consider data about the general elections for the years 2002, 2006, 2010 and 2014 (with disputes for the positions of President, Vice President, Governor, Deputy Governor, Federal Deputies, State Representative and Senators) and the local elections of the years of 2004, 2008, 2012 and 2016 (with disputes for the positions of Mayor, Deputy Mayor and Councilman).

3 Network Power Metrics and Linked Open Data

We propose a framework, presented in Fig. 1, to construct and explore a knowledge graph to investigate connections between agents involved on slavery labor and politicians, evaluating their power and influence. The framework takes as inputs datasets of different data sources (in different formats), ontologies and controlled vocabularies to outputs a knowledge graph, semantically enriched, following the Linked Open Data standards.

This knowledge graph enables a number of interesting analyses and explorations, such as: (i) Social Networks Analysis (SNA) metrics can be used to analyze different aspects of the network, e.g., power and influence of the agents; and (ii) different explorations combining search, navigation and SPARQL queries that can be useful to understand how the enslaving agents are connected to the Brazilian politicians.

---

[1] CPF is the fiscal identification of Brazilian citizens. CNPJ is the fiscal identification of Brazilian companies.

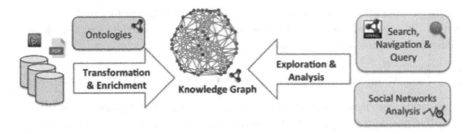

Fig. 1. Overview of our approach framework

3.1 Metrics to Measure Power

This research applied metrics based in Manuel Castell's Network Theory of Power described in [10] to analyze the power hold by slavery agents. These metrics use concepts from sociology that can be summarized in three aspects: (i) nodes are connected by the exchange of resources that are valuable for the network; (ii) when a node has the capability to change network rules it is a powerful node; (iii) the most crucial ability in a networked society is the ability to bridge two or more different networks. The set of metrics measures which is the most powerful side in an exchange of resources. For this is used primarily the idea that if one side has alternative options to exchange its resources it is strengthened in the negotiation, according to Castells a key point for obtaining power: the more and better alternative options the agent possesses to exchange resources, less important a specific exchange is for him and less dependent on this relationship he is. The power of a node can be expressed as the sum of the power accumulated throughout its network of resource exchanges.

But what is a node, what is an edge and how to assign the weight to an edge? These fundamental decisions can significantly alter the outcome of any measurement or topological process that runs over a network. Although deeply rooted in our contemporary minds, the concept of networking is always an abstraction, with no backing in the real world. Not even a very concrete network, such as a computer network, can escape inaccuracies if we try to specify exactly what the nodes are (the whole computer, only the network card, a connector?) and what connects them (antennas, protocols, software?). The hypotheses about the nature of power used to build the metrics inherit sociological concepts and assume that the modeled network has two characteristics: (i) nodes are social agents, that is, people or organizations of people at their most diverse levels, such as political parties, companies, councils entrepreneurs, foundations, social organizations, etc.; and (ii) the connections between these agents take place through exchanges of value resources to the network carried out in the real world. What is a value resource will depend on the network you are building, and a most obvious example would be money transfers between agents. Other types of resource transfers could be used as edges, such as: environmental licenses, transfer of properties, advertising, services performed, etc. The edge has origin and destination and therefore is directed, following the same direction of the flow of resources. The weight of the edges will indicate the volume of resources transferred.

3.2 Building a Network

Through the research sources described in Sect. 2.1, all 23 editions of the Dirty List of Slavery were recovered since its creation in 2003 until January 2018. Figure 2 shows the evolution of the number of operations and Fig. 3 the evolution of the number of rescued workers through time.

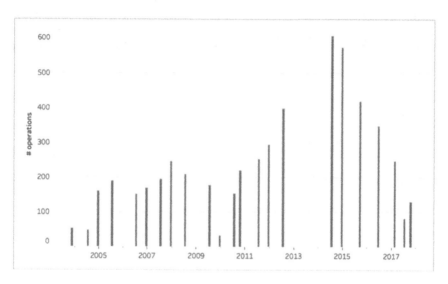

Fig. 2. Number of operations in each edition of the Dirty List of Slavery (2004–2018).

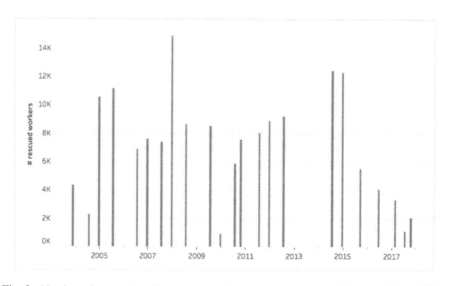

Fig. 3. Number of rescued workers in each edition of the Dirty List of Slavery (2004–2018).

It was possible to identify each operation and employer of the list (removing the ambiguities and cleaning up the CPFs and CNPJs), reaching the final number of 1,665 operations, involving 1,553 employers and 39,343 employees.

To perform the power analysis, these employers were interconnected with politicians, using campaign donations as edges, modeling a network that follows the process discussed in Sect. 3.1. Thus, each individual or legal agent present in the Dirty List of Slavery represents a node, as well as each politician and political parties that have connections to them. The network is formed by edges representing resource transfers in campaign donations. A network view as presented in [11] can be seen in Fig. 4. This view was generated using software Gephi with the SigmaJS plugin. This network can be navigated and explored in www.trabalhoescravo.info.

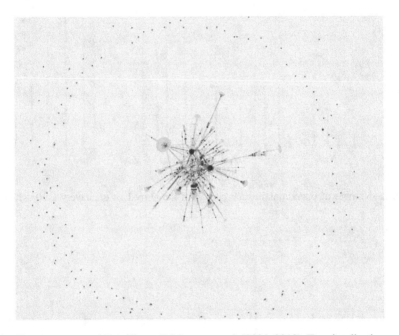

Fig. 4. Slavery agents and Brazilian politicians network (2004–2018). For visualization purpose the size of the nodes is proportional to the volume of resources transferred through it. Brown nodes represent slavery agents, yellow nodes represent politicians and green nodes represent political parties. (Color figure online)

Table 1 shows some traditional SNA metrics [12] to give a glimpse of the network structure and size.

Table 1. SNA metrics of the network

Measure	Value
Number of nodes	1,853
Number of edges	2,156
Diameter	7
Average path length	2.341
Modularity	0.546
Weakly connected components	543

3.3   Linked Open Data and a Knowledge Graph to Explore Slavery Data

The simplified network shown in the previous section and used to calculate power metrics does not include the publication of all information captured by the research. Questions about when the donations where made, to what role the candidate was applying and many others cannot be answered. Using Linked Open Data (LOD) modeling, a knowledge graph was created containing all the information and generating a rich dataset for future research. These data allow query, navigation and interconnection with other datasets using W3C standards, enriched with metadata that allows inference by machine and future quality surveys on the data. The interconnection between data from different sources is a fundamental point to add value and gain knowledge of the data [13]. Berners-Lee addressed data interconnection problem with the paradigm known as Linked Open Data (LOD) or Semantic Web [14]. The Semantic Web can be considered a worldwide movement for data exchange, which includes pressure to use patterns and to break walls that prevent or hinder broad access to data, especially with a focus on the use of vocabularies and ontologies for understanding the meaning and context of the data source.

The best practices recommended by the W3C for publishing data are the use of HTTP URIs (Uniform Resource Identifiers) as identifiers of what is being described; the provision of standardized information when the URI is referenced; and binding to other URIs, so that related information is discovered. The W3C also proposes the use of RDF (Resource Description Framework) as a standard for graph construction in the Semantic Web. RDF graphs are composed of *subject-predicate-object* triples. This structure allows the use of vocabularies and ontologies to describe the data, so that it is possible to provide software that will process the data according to their meaning and support inferences.

Ontologies are structures and formal rules that represent a worldview of a portion of reality that guide the generation of data. They allow the reconciliation between different models and data from different sources, since the intentions are explicit in the metadata. This modeling allows implicit rules to be understood and processed. For example, if we have an ontology whose rules determine that an application for a role can only be made by a person, and the data states that someone has applied, the computer can understand that this agent is a person.

Vocabularies are less strict and formal than ontologies, but they also describe data using standard terminologies. Some widely used ontologies and vocabularies support the description of billions of objects worldwide. Some examples are FOAF, Schema.

org and DBpedia. The latter has the same name as a multi-domain dataset from Wikipedia, today the most popular Semantic Web data source in the world, interlinked with numerous government data, libraries and other thematic databases.

Integrating information from different sources is costly in terms of time and money, and the basic idea of the Semantic Web is to create efficient ways of publishing information in distributed environments, reducing independent efforts through the use of widely disseminated standards [15]. The modeling effort required for the publication of data in the form of Linked Open Data should be seen as a step towards unclogging data access, since, by modeling the data to function independently of the context of a specific application, we are preparing them for any use. To model the research data concerning politicians and their campaign donors, the POLARE ontology presented in [16] was used. One of the challenges that was immediately posed when selecting the data sets to be worked on is the correspondence between instances. Connecting search data with DBPedia for example would have to use some matching method, since the attribute for a possible direct link - the CPF - is not present in DBPedia. We decided not to carry out this interconnection in the scope of this work, but more sophisticated methods of correspondence, such as that proposed by [17] should be applied to overcome this obstacle and enrich the results of future research.

As a final result of this research, an RDF dataset was generated that can be queried using SPARQL queries [18] and the publication of this dataset in DBpedia has been planned.

4   Results: Connections Between Politicians and Slavery Agents

The data built network revealed 191 political candidacies from employers who featured on some edition of the Dirty List of Slavery for which they received about 21 million Brazilian Reais in campaign donations. In addition, employers on the Dirty List have donated about R\$ 48 million to other candidates evidencing their struggle for power and influence. Table 2 shows the biggest campaign donors to slavery agents, Table 3 the top political parties who received money from donors that in some point of time were present at the Dirty List of Slavery; and Table 4 the individual politicians in the same situation.

Table 2. Top campaign donors according to amount donated (2004–2016).

Donor	Amount donated
CCM - Construtora Centro Minas Ltda.	R\$ 8.443,231,00
Construtora Central Do Brasil S/A	R\$ 8.333.755,00
Tratenge Engenharia Ltda.	R\$ 3.698.264,00
Usina Siderúrgica De Marabá S/A	R\$ 1.985.737,00
Emival Ramos Caiado Filho	R\$ 1.339.000,00
Laginha Agro Industrial S/A	R\$ 1.254.976,00
Jose Essado Neto	R\$ 1.250.334,00
Siderúrgica Do Maranhão S/A	R\$ 1.225.000,00
Valdir Daroit	R\$ 1.074.040,00
Jair Correa	R\$ 963.582,00

Table 3. Top political parties according to the amount received in donations (2004–2016).

Donor	Amount donated
PSDB	R$ 3.943.541,00
PMDB	R$ 2.451.457,00
PR	R$ 1.827.462,00
PT	R$ 1.435.113,00
PSB	R$ 978.700,00
DEM	R$ 633.916,00
PPS	R$ 534.598,00

Table 4. Top politicians according to the amount received in donations (2004–2016).

Donor	Amount donated
Sergio Ramos Caiado	R$ 1.339.000,00
Jose Essado Neto	R$ 1.202.500,00
Joao Jose Pereira de Lyra	R$ 933.476,00
Mauro Luís Savi	R$ 923.740,00
Jair Correa	R$ 863.582,00
Cassio Rodrigues da Cunha Lima	R$ 800.000,00
Alcides Rodrigues Filho	R$ 635.000,00
Erik Augusto Costa e Silva	R$ 572.840,00
Aécio Neves da Cunha	R$ 567.799,00
Antônio Carlos Bacelar Nunes	R$ 565.065,00
Agenor Rodrigues de Rezende	R$ 520.936,00
Hélio Calixto da Costa	R$ 500.000,00
Samuel Moreira da Silva Júnior	R$ 500.000,00
Edison Lobão Filho	R$ 500.000,00

The application of power metrics can also show that some interests are not directly linked to the amount of money given in the electoral donation. Table 5 shows the most powerful politicians according to the metric as an example of this aspect analysis. The used metric connect campaign donors to their politicians and parties in order to find which the most powerful part in the deal is, and that's the reason it highlights some names not revealed if only the amount is considered.

The generated knowledge graph supports search, navigation and queries, leveraging the exploratory investigation of the data. As the metadata are connected to each data item, the graph is easy to understand and the modeling decisions are explicit.

As an example of the potentialities of the knowledge graph, Figs. 5, 6 and 7 show a cross-data navigation generated by loading the graph in the GraphDB software. The RDF dataset and ontologies used can be downloaded from www.trabalhoescravo. info.

Table 5. Top politicians according to the power metric (2004–2016).

Agent	Power metric
Jose Pedro Goncalves Taques	1459.118502
Ronaldo Ramos Caiado	1380.090807
Altineu Cortes Freitas Coutinho	1274.505642
Blairo Borges Maggi	933.765730
Heraclito de Sousa Fortes	917.249407
Antonio Carlos Arantes	748.522568
Renato Barbosa de Andrade	717.862038
Bernadete Ten Caten	476.629796
Cassio Antonio Ferreira Soares	469.183049
Jovair de Oliveira Arantes	366.153015

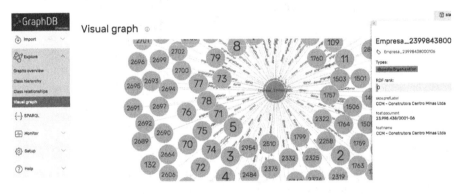

Fig. 5. Results for a search in the semantic graph with the term "Construtora Centro Minas"

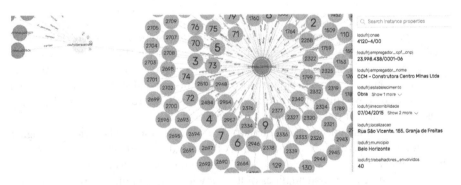

Fig. 6. Expansion of the node representing the operation carried out by the Labor Department of Justice that caught the use of slave labor by Construtora Centro Minas

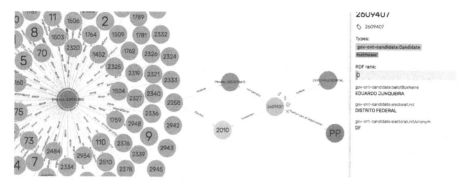

Fig. 7. Expansion of a node representing a campaign donation made by Construtora Centro Minas to a politician

Through SPARQL queries it is possible to generate an electoral dossier of each of the employers caught using slave labor, as can be seen in the timeline generated as an example and shown in Table 6. The employer used in the example, Jose Essado Neto, appeared in the Dirty List of Slavery in July/2014 and December/2014 caught using slave labor in 2007, when 181 workers were rescued. Although the values in the Table may seem small, they are sometimes "estimated", what possibly mean they are underestimated by the politician who presented the report.

Table 6. Electoral dossier of employer Jose Essado Neto

Year	Activity
1998	State Deputy Candidacy (PMDB), State: GO. Elected
2000	Mayor Candidacy (PMDB), City: Inhumas/GO. Elected
2002	Donated to state deputy candidate Divino Rufino da Silva (PMDB). State GO. R$ 8,910, money donation
2004	Mayor Candidacy (PMDB), City Inhumas/GO. Not Elected. Donated R$ 380,000 to his own candidacy, money donation
2006	State Deputy Candidacy (PMDB), State GO. Elected. Donated R$ 510,000 to his own candidacy, money donation. Received R$ 50,000 from Centrocouros Inhumas Ltda, check donation
2008	Donated to mayor candidates: R$ 150 to Victor Leonardo Lima Soares (PMDB). City Adelândia/GO, R$ 100 to Ana Paula Gonzaga Souza (PMDB). City Aruanã/GO, R$ 1,760 to Waltenir Peixoto de Miranda (DEM). City Avelinápolis/GO, R$ 100 to Luiz Carlos de Sousa (PMDB). City Caturaí/GO, R$ 180 to Americo Osorio Santos e Silva (PMDB). City Damolândia/GO, R$ 120 to Paulo Cesar Raye Aguiar (PTB). City Faina/GO, R$ 500 to Célia de Morais Marques (PMDB). City Guaraíta/GO, R$ 12,000 to Dioji Ikeda (PMDB). City Inhumas/GO, R$ 500 to Luiz Carlos da Silva (PMDB). City Mozarlândia/GO, R$ 300 to Edilson Galdino Rocha (PMDB). City Taquaral de Goiás/GO. Estimated donations Donated to councilman candidates: R$ 509 Deny Leles Rosa (PMDB). City Morro Agudo Goiás/GO. R$ 140 to Egnaldo José de Carvalho (PP). City: Araguapaz/GO. R$ 40 to Murilo Moreira Teles Valim (PMDB). City Caturã/GO. R$ 80 to Raimundo Soares Amorim Neto (PSDC). City Goianira/GO. R$ 140 to Ildebrando Potenciano Neto (PR). City Mozarlândia/GO. R$ 20 to Osvair Jose dos Reis (DEM), Denival Ribeiro de Souza (DEM), Osmando Rosa Vicente Souza (DEM),

(continued)

Table 6. (*continued*)

Year	Activity
	Valdeir Pereira de Oliveira (DEM), Joao Batista Costa (PMDB), Auxiliadora Santos Araujo (PMDB), Joao Lacerda de Souza (PMDB), Coraci Maria Delurde, Adao Domingos da Paz (PMDB), Maria Auxiliadora Serra (PMDB), Alfredo Herwig (PMDB), Celio Lopes da Silva (PMDB), Jose Maria Dantas (PTB), Junior Carlos dos Santos (PT). City Matrinchã/GO. Estimated donations Donated to political parties: R$ 2,040 to PMDB, R$ 1,685 to PP, R$ 986 to DEM. Estimated donations
2010	State Deputy Candidacy (PMDB), State: GO. Elected. Donated R$ 742,500 to his own candidacy. Money donation. Donated R$ 15,000 to his own candidacy. Estimated donation
2014	State Deputy Candidacy (PMDB), State: GO. Not Elected Donated to state deputy candidate Adenilson Pessoni (PMDB). State: GO. R$ 5,000, check donation
2016	Donated to councilman candidates: R$ 194 to Laryssa Bueno Souza (PMDB), R$ 130 to Cristiana Rodrigues Ribeiro (PMDB). R$ 140 to Dilmário Alves Pereira (PMDB). R$ 150 to Laila Rafaela Alves Borges (PMDB). R$ 180 to Clodoaldo Costa Ferreira (PMDB). R$ 320 to Itubes Vanderlino Brito (PMDB). City: Inhumas/GO. Estimated donations

It is to be noted the geographic concentration of donations: the employer seeked for power donating for many candidacies for deputies, mayors and councilmen in the same state, GO. This is a clue for the reason some cities seem to be doomed by slavery episodes. The state of Maranhão (MA) is the biggest source of enslaved workers according to the Digital Enslavement Observatory Smartlab[2]. One small municipality from this state, named Codó, is the second city in number of enslaved residents in the country. The knowledge graph can also answer geographically based queries to understand this is not a coincidence. Table 7 shows the political activity of enslavers in Codó as example, also generated by SPARQL queries in the graph.

Table 7. Electoral activity of enslavers in the city of Codó/MA

Year	Activity
2000	Mayor candidacy José Rolim Filho (PDT). Not Elected
2004	Mayor candidacy José Rolim Filho (PV). Not Elected Donation from Francisco Antelius Servulo Vaz to mayor candidate Benedito Francisco da Silveira Figueiredo (PFL). Donation Value: R$ 18,000.00
2008	Mayor candidacy José Rolim Filho (PV). Elected Donation from political party PV to mayor candidateJosé Rolim Filho (PV). Donation Value: R$ 1,005,266.90 Donation from Sergio Marcos de Assis to mayor candidate Ildemar Gonçalvez dos Santos (PSDB). Donation Value: R$ 15,000.00
2012	Mayor candidacy José Rolim Filho (PV). Elected Donation from political party PV to mayor candidateJosé Rolim Filho (PV). Donation Value: R$ 600,407.38

(*continued*)

---

[2] Partnership between ILO and Brazilian Labor Department of Justice.

Table 7. (*continued*)

Year	Activity
	Donation from Gráfica e Editora JM Ltda. to mayor candidate José Rolim Filho (PV). Donation Value: R$ 6,850.00 Donation from Batista e Silva e Cia Ltda. to mayor candidateJosé Rolim Filho (PV). Donation Value: R$ 2,000.00
2014	Donation from José Rolim Filho to federal deputy candidate Ricardo Archer (PSL). Donation Value: R$ 1,500.00
2016	Donation from José Rolim Filho to coucilman candidate Francisco Nagib Oliveira (PDT). Donation Value: R$ 880.00

5   Conclusion

The contributions of this work can be summarized as follows: (i) an unpublished dataset containing all editions of the List of Slave Labor in standard LOD format was generated; (ii) the enslaving agents were connected to the Brazilian politicians, and a standard LOD knowledge graph was created containing the network; (iii) on this network, metrics were used to measure the power of these agents over public agents, providing an overview of the power of these criminals and making possible the future use of this data both for the generation of new researches and for the information of activists fighting for eradication of slave labor in Brazil.

Regarding future work, the inclusion of degrees of kinship in this knowledge graph can reveal new connections between criminals and public power. It is crucial to create user-friendly and publicly available visualization tools for the general public, so that citizens, activists and journalists can act as inspectors of public power and allies in the fight against contemporary slave labor. This crime, globally condemned by various international agreements is practiced by a greedy elite, who uses the suffering of these slaves to increase their economic power, uses economic power to gain political power, and uses political power to perpetuate exploitation. Crime that is only an example of the Brazilian segregation that structurally oppresses a portion of its citizens, against whom all violence is naturalized; against whom all exploitation is permitted; for whom the centuries of slavery insist on not ending.

Acknowledgments. The authors would like to thank CNPq and FAPERJ for partially supporting this work.

References

1. Bales, K.: Blood and Earth: Modern Slavery, Ecocide, and the Secret to Saving the World. Spiegel & Grau, New York (2016)
2. Datta, M.N.: Using big data and quantitative methods to estimate and fight modern day slavery. SAIS Rev. Int. Aff. **34**(1), 21–33 (2014)
3. NGO Repórter Brasil Homepage. http://reporterbrasil.org.br/. Accessed 21 April 2018

4. Religious Organization Comissão Pastoral da Terra Homepage. https://cptnacional.org.br/. Accessed 21 April 2018
5. Obama, B.: Memorandum for the heads of executive departments and agencies. Pres. Stud. Q. **39**(3), 429 (2009)
6. Brazilian Open Government Data Portal (Portal de Dados Abertos). http://dados.gov.br/. Accessed 5 Mar 2018
7. International Labour Organization (ILO) Homepage about slavery. http://www.ilo.org/global/topics/forced-labour/lang–en/index.htm. Accessed 25 April 2018
8. Viana, M.T.: Trabalho escravo e "lista suja": um modo original de se remover uma mancha. Organização Internacional do Trabalho, Brasília (2007)
9. Brazilian Election Data Portal (Repositório de Dados Eleitorias) Homepage. http://www.tse.jus.br/hotSites/pesquisas-eleitorais/resultados.html. Accessed 5 Mar 2018
10. Campos, M.L.M., Oliveira, J.: Métricas para análise de poder em redes sociais e sua aplicação nas doações de campanha para o senado federal brasileiro. In: Proceedings of XXXVII Congresso da Sociedade Brasileira de Computação, São Paulo, pp. 544–554 (2017)
11. Fruchterman, T., Reingold, E.M.: Graph drawing by force directed placement. Softw.: Pract. Exp. **21**(11), 1129–1164 (1991)
12. Attard, J., Orlandi, F., Auer, S.: Value creation on open government data. In: Proceedings of IEEE, 49th Hawaii International Conference on System Sciences (HICSS), pp. 2605–2614 (2016)
13. Easley, D., Kleinberg, J.: Networks, Crowds, and Markets: Reasoning about a Highly Connected World. Cambridge University Press, New York (2010)
14. Berners-Lee, T.: Linked Data - Design Issues. W3C. http://www.w3.org/DesignIssues/LinkedData.html (2006)
15. Bauer, F., Kaltenbock, M.: Linked Open Data: The Essentials. Mono/Monochrom, Vienna (2011)
16. Laufer, C., Schwabe, D., Busson, A.: Ontologies for representing relations among political agents. arXiv preprint arXiv:1804.06015 (2018)
17. Araújo, S., Tran, D., De Vries, A., Hidders, J., Schwabe, D.: SERIMI: class-based matching for instance matching across heterogeneous datasets. IEEE Trans. Knowl. Data Eng. **27**(5), 1397–1440 (2015)
18. Segaran, T., Evans, C., Tayloy, J.: Programming the Semantic Web. O'Reilly Media Inc., Sebastopol (2009)

Collaboration and Crowdsourcing

CidadeSocial: An Application Software for Opportunistic and Collaborative Engagement of Urban Populations

Ana Clara Correa[1], Eliel Roger[1], Tiago Cruz de França[1,2(✉)],
José O. Gomes[2], and Jonice Oliveira[2]

[1] DECOMP, Federal Rural University of Rio de Janeiro, Seropédica, RJ, Brazil
tcruz.franca@gmail.com
[2] PPGI, Federal University of Rio de Janeiro, Rio De Janeiro, RJ, Brazil
jonice@gmail.com

Abstract. The combination of mobile devices and easy access to the Internet enhances the spread of information coming from our day-to-day lives. People share all types of information such as events, opinions and problems in urban areas. Also, they can act as sensors by monitoring and sharing information about the demands made by inhabitants for urban changes on social media. Moreover, these platforms allow people to support each other - even strangers - through questions and answers, recommendations and indications. However, much of this kind of information is lost. Even social media (e.g. Facebook and Twitter) limit the spread of this flow of information up to the network borders, because their focus is on a relationship network. Consequently, the information does not reach people outside these networks. This paper describes an application software - named CidadeSocial – that allows inhabitants to share information according to their common interests. The application software exploits the geospatial location of the users to create a temporal social network, provides recommendations based on their profile and uses a gamification approach to encourage user engagement. Thus, we argue that CidadeSocial is a tool with potential to serve as an interface for engagement in improving the day-to-day life of cities and their inhabitants.

Keywords: Opportunistic collaboration · Mobile computing ·
Gamefication · Communication · Exchange of information

1 Introduction

People are members of different social networks (e.g. group of friends, coworkers and relatives) [1]. Nowadays, people use various social media to interact and build their social networks, such as Facebook[1], Linkedin[2] and Twitter[3]. However, since these media do not focus on city dynamics people search for information beyond the boundaries of their social networks. For example, there was a much higher number of

---

[1] https://www.facebook.com.
[2] https://www.linkedin.com.
[3] https://twitter.com.

© Springer Nature Switzerland AG 2019
J. Oliveira et al. (Eds.): BiDU 2018, CCIS 926, pp. 141–155, 2019.
https://doi.org/10.1007/978-3-030-11238-7_9

cases of yellow fever (a disease that had already been controlled and practically eradicated many years ago) throughout Brazil in the last two years (2017 and 2018). As this is considered an emergency situation, immunization of the population was made mandatory to control the epidemic. Although information concerning vaccinations was disseminated by the social governmental media, some citizens had questions such as: Who can be vaccinated? Who should not receive the vaccine? Under which situations should I have a clinical check-up before the vaccine? In such cases, people tried to get answers from their networks, but often without success.

In this case a lack of information or worse misinformation can cause problems in the health system, bring panic to the local population and even create problems for health workers (i.e., people go to the local hospital to get information, and interrupt the services and employees). These doubts and answers (the information) should not be restricted to a "network of friends" or to just another social media, instead they must be available for everyone on a more open "social media".

Moreover, people have to interact to overcome common problems. For instance, if a specific region of a city provides a poor public transport service, the life quality of the inhabitants of this region will be lower than another region with better transport. The people affected who want to complain about such problems need to communicate with each other. However, this communication is not always easy; moreover, these individuals probably do not even know each other. Several other problems arise in big urban centers, such as: lack of information about local events or the availability of public services nearby; access of the population to civil services and their managers; and lack of transparency related to city problems in general.

These examples are common in cities, but even so there is still a lack of tools to support any spontaneous collaboration. An opportunistic collaboration is the one that starts with an unintentionally encounter where one individual takes advantage of a situation to talk and discuss some problem with another individual [2]. Through such spontaneous actions and collaborations focused on common interests opportunistic networks can be set up [2].

This paper presents the CidadeSocial (a combination of two Portuguese words that mean city and social, respectively), an application software for opportunistic communication using mobile devices. The idea of this application software is to support citizens in their day-to-day requirements in cities, describing and georeferencing problems in a collaborative way. CidadeSocial employs gamification strategies to attract users and spread awareness. Also, it uses geospatial information and user profiling to send custom information to users.

The rest of this paper is organized as follows: Sect. 2 presents the related works; in Sect. 3 we describe the CidadeSocial (functions and architecture); Sect. 4 gives some implementation and examples of use; and Sect. 5 has the final considerations and future works.

2 Related Work

A city can be defined as an "urban agglomeration", a contiguous stretch of a built-up area [3]. The dynamics of cities can be related to any relationship between the geographical space and its inhabitants. Examples of dynamics are: commuting; migratory flows; the formation of dynamic events such as the formation of protests; and changes in population density due to housing or industrial developments in the city, among others. Each city has its unique characteristics, dynamics and problems that affect the citizens.

Another related concept is "smart cities" which is related to the potential use of mobile devices, web technologies and citizen participation in the day-to-day life of cities [4]. The authors of [5] pointed out the need to provide appropriate interfaces for citizens to engage actively in building an intelligent environment. Here the aim of CidadeSocial is to provide the means for information exchange using mobile devices, and to function as a tool for engagement of urban dwellers, as well as to become in time a source of information for a city.

Several papers have analyzed the semantics added by georeferenced data to social media messages [6, 7]. In [8] the authors analyzed data from Foursquare[4] to predict crimes in New York. Although these authors used georeferenced data, they did not provide an appl to get data from individuals as is proposed in this paper. CidadeSocial aims to get geo tagging data (coordinates) of users, places and organizations. These geo referenced data enable the search mechanism to show the topics published near the geographic location of the user. In other words, the posts are sorted according to how close the user is to the geo localization in the posts.

An Instagram[5] and Foursquare evaluation was made in [7]. The authors wanted to investigate these media as being participatory sensing networks, where data from sensors (e.g. GPS and barometers, etc.) captured by the mobile devices are attached to the user-provided data [2]. Then, they described Instagram as being a media focused on cultural issues, and Foursquare as being a participatory sensing network for the definition of way-finding. Thus these authors pointed out the usefulness of these media as tools to understand both the dynamics of displacement (by tourists, for example) and the acquisition of information concerning cultural events. The aim of CidadeSocial is in communication centered on user interests (which are related to city dynamics) and not only on cultural or issues of location as Instagram and Foursquare.

Applications such as Google Maps[6] and Waze[7] focus on traffic and location in urban spaces. They deal with traffic and way-finding information by exploring user collaboration. But they do not focus on user communication as CidadeSocial does. CidadeSocial uses maps to add value to the communication and collaborate with functions.

---

[4] https://pt.foursquare.com/.

[5] https://www.instagram.com.

[6] https://www.google.com/maps.

[7] https://www.waze.com.

Foursquare and Waze use gamification in order to motivate their users. However, their goal with gamification is different to that of CidadeSocial. The primary goal of CidadeSocial is to disseminate information related to the day-to-day routines of the city dwellers, which is something more than just focusing on the traffic or registering places where a user passes by. However, to increase user commitment, gamification techniques are also used here in CidadeSocial [9]. As a reference, we adopted [10] where gamification included gamefulness elements (game experience), gameful interaction (tools and objects and contexts that will be part of the user experience) and gameful design (process of elaboration of that experience) in both context and purpose.

3   CidadeSocial

CidadeSocial is an application software to spread information using mobile devices. The propagation of the information requires user collaboration which occurs when they exchange information. The application includes a recommendation system (to build the user timeline), an offline mode (to make it usable even if temporarily without Internet access), a gamification approach (to encourage its use), a search service (exploring georeferences and grouping messages by interests), a topic (a message initiating a thread communication) publishing service, a topic comment service, and a collector of user geo coordinates using their GPS devices. The CidadeSocial app is fully described in this section.

3.1   From Universities to Cities

In the early versions of this project the goal was to promote information exchange inside university campuses. Some academic communities have inherited historical communication problems, such as (1) difficulties for getting information on university buildings, (2) getting the student body to take notice of both official communicates and (3) events which are happening at the University [11]. Some Brazilian Universities are as big as small cities and so their campus is manage with the same organizational structure as a city (i.e., mayor, security rules, own transportation system and other public services). Now, we have expanded the project to be adapted to big urban centers, with more functions. This was possible due to the similarities between universities and cities.

3.2   Cidade Social Feature Description

The communication on CidadeSocial is based upon interests. The interests are defined as the strategy to link users, their interests and message content. All information is associated with an interest. Therefore, users receive preferably information of their interest.

The interests of the user are defined during the setting up of an account and can be updated anytime. Table 1 shows the list of predefined interests. Differently from the more common social media (such as Facebook), the communication (the content of messages) in CidadeSocial comes before a person-to-person relationship. In other words, the main purpose of CidadeSocial is to enable the spreading of information trying to reach everyone and anyone interested in a subject (or interest) even if they are strangers to each other.

The information is centered on topics, which are messages to initiate a communication thread (such as in a web forum). Each topic must be associated to one or many interests (see Table 1). The interest constraint enables the linkage between topics and users. All topics are public and their dissemination is not limited to followers. Instead, they can reach anyone who has an interest in such a topic. Any user can comment on a topic. Comments are replies to a topic.

The messages (topics and comments) are associated with the geolocation (geography coordinates) of the user when published. The coordinates can be used to give priority to show that this message user is close to this location. In addition, they can be used to aggregate value to messages during a possible data analysis or to power other analyzes.

Table 1. List of Interests.

Interests			
Sport	Recreation	Complaint	Denouncement
Praise	Disclosure	Art	Music
Theater	Entertainment	Event	Infrastructure
Service	Security	Crime	Violence
Trade	Employment/work	Health	Social work
Mobility	Streets/avenues	Transport	Food
Politics	Elections	Government	Legislative
Town hall	Municipal council	Judiciary	Protest

Topics can be associated to a place or an organizational unit. Places are in areas or regions that can be identified. Examples of places include, but are not limited to: streets, squares, bus stops, subway or train stations, beaches, tourist spots, etc. An Organizational Unit where a unit represents an organization with multiple units (or departments). For example, the City Hall of Rio de Janeiro (organizational unit) is in Floriano Square (place), Cinelândia, at Coordinates $-22°54'36.175''N$ $-43°10'36.108''E$. Place and organizational units are registered in the database (by administrators) each one containing a unique name and their location (geography coordinates are also used, for example, to pin places and organizations on a map). The main purposes are to relate places and organizations to topics, show them in a web map service (improving the visualization service) and enable the location of place in a region (e.g. where the square or hospital in that neighborhood is).

Users can evaluate topics, places and organizational units by giving a note from 0 to 5 (in a Likert scale[8]). A user can only evaluate an item (place, organization or topic) once. The average value is presented to all users as soon as anyone sees the place,

---

[8] https://en.wikipedia.org/wiki/Likert_scale.

organization or evaluated topic. This evaluation is used when ranking topics during the building of a timeline and it provides feedback about places or organizations (e.g., how people evaluate some place or organization unity?).

The user timeline is made up of topics sorted by interests, published near the actual user geo-coordinates at the time. These rules were adopted as the first approach; however, they can be adapted in the future to improve this service if necessary. The timeline is the first screen that the user sees after login. It represents a copy of the data in the server and is custom built by the user, and there is a recommendation system that takes care of these user interests. Another screen is the home screen (visualization of topics) in which all the topics of the user are presented (topics crated by the user him/herself).

Access to the Internet may be lost during the day due to location for example; however there is an offline mode which allows the use of some functions without the Internet, and therefore there is a client sync component. In the offline mode a user can create a topic, see their timeline (limited to the content already there), evaluate topics, and comment topics available since the last synchronization. Synchronization always occurs when the Internet connection is or becomes available. The Mobile device automatically triggers an update with the server each 30 s in order to assure that the user will have access to updated information. The data produced, when offline, are sent to the server as soon as the connection to the Internet returns. The reason for the offline mode may arise from possible problems of mobile Internet services such as areas of coverage. Moreover, the user can restrict access to the Internet only when Wi-Fi is available.

The user localization refers to his geo coordinates whenever he uses the system. The coordinates will be obtained from the GPS when it is available. The coordinates are used when a user searches for information. The information will be sorted according to how close the user is to the place of publication of the messages returned in the search. Other types of searches such as search by interests, places or organizational units are also available.

A gamification mechanism was developed. It is described in detailed in [9]. Its goal is to motivate users, by reward mechanisms. Users earn points and achieve levels according to their actions in the application. The ranking which is based on the users progress is available. This score is used as a metric of credibility for his posts and comments.

3.3   The CidadeSocial Architecture

CidadeSocial is a client-server model distributed system. We adopt the Representational State Transfer (REST) architectural style to design the application over HTTP. All communications are made in security channels over HTTPs. The OAuth2[9] protocol was used for user authorization control. Figure 1 illustrates the distributed model of the CidadeSocial.

Essentially, both client and server are made up of components, where each has it own interfaces that allow them to communicate with each other. On the client side, the

---

[9] https://oauth.net/.

components provide user interfaces that will be related to user operations - sync, log in or sign in. On the other hand, the server side will prepare and provide end points for web requests.

The CidadeSocial Client. Figure 2 presents the tasks of the client for the most common CidadeSocial functions. Figure 3 presents the diagram of the user components that implements these functions.

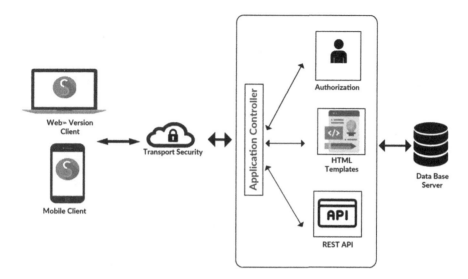

Fig. 1. Overview of the CidadeSocial client-server.

The User Manager component provides the user registration, the selection of user interests, user authentication and password recovery services. After the login, the user receives the OAuth credentials. These credentials are what ties a requisition to the author, which allows the server to identify the author and verify if he/she is authorized to access some resource. The AssociationPost and CredentialsMessage interfaces are responsible for linking the post and user credential. This component depends on the DataManager interface.

The Data Manager stores all client data which include topics for timeline (with comments, locations, etc.), user data and user topics (including unpublished data). The Data Manager has an interface with the same name DataManager and depends on the SynchronizeData interface.

The Sync is responsible for keeping the client and server synchronized. It keeps a copy of the data which is composed of recommended topics for the timeline, the user topics, etc. If there is a new topic, comment or evaluation, this component is responsible to update the server. Besides, the Sync supplies the SynchronizeData interface and depends on the Connection interface. The sync was not represented in the activity diagram but it has the prominent role of making everything work.

The Connecting Manager component is responsible for sending and receiving messages. It handles all the communications established with the server. All messages

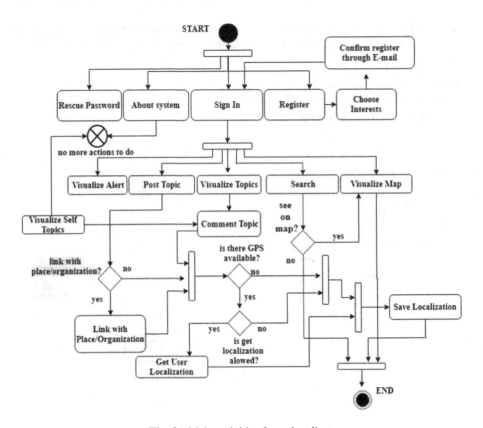

Fig. 2. Main activities from the client.

sent pass through it and require the user OAuth credentials (obtainable from the CredentialsMessage interface). It also checks the Internet (through the Connection interface) connection.

The user location is obtained by the Location Manager. User permission and signal identification (or GPS if available) are controlled by this component. Other client components acquire this data through the GetPosision interface.

The Place Manager and Organization Manager basically identify and maintain the place and organization information considering how close they are to the user. They supply the GetPlaceData and OrganizationData interfaces respectively.

The Evaluation Manager controls the evaluation service. It supplies the Interface Manager Evaluation. Users can evaluate topics, places and organizations.

The association of points and experiences depend on the activity performed. This information is associated with the requests to be sent to the server. The component Gamification-engine provides access to this function through the GameTask interface.

The Search component is responsible for providing ways to find topics according to user interests. It also matches user location and the geography coordinates associated to topics. The components depend on the GetPosition interface to obtain the user location.

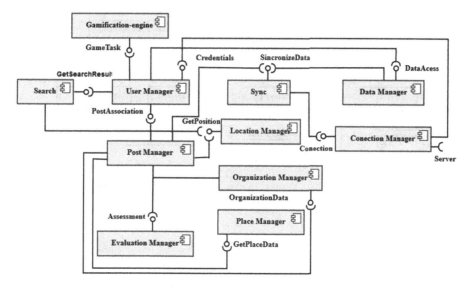

Fig. 3. Main client components

Finally, the Post Manager component is responsible to send all messages (topic, comments and alerts) to the server (for publications). All posts are associated with the user who published them (through the AssociatePost interface). The alerts are only available to administrators. This decision was made because the alerts, which are broadcasted to all users, could be used in an inappropriate way (by an unauthorized individual). An example of an alert is: "the subway is not working, because the employees are on strike).

The CidadeSocial Server. The server makes a web API available for all resources. All the contents of the CidadeSocial pass through the server. In other words, all messages (or evaluations) published by clients are sent to the server which registers the activities and stores the data in a database. The diagram in Fig. 4 represents the main components, interfaces and dependencies of the server. Some components have similar names to those belonging to the CidadeSocial Client, but their objectives are different.

There is a software component in the server responsible to provide access to all other components through the web, which is named CidadeSocialAPI, and is based on the REST principles. The CidadeSocialAPI receives all HTTPs requests for each service (resource) provided by the server. All resources have at least a URL as defined in REST principles. Table 2 presents examples of some URLs.

All communications depend on the Authorization component to obtain the credentials of the client application. This component is responsible for managing the client credentials. The OAuth2[10] is used to authorize any access. The component defines the client authorization using their credentials (which are associated with their permission). The client has to send the credentials for each request. The application is stateless, so

---

[10] https://tools.ietf.org/html/rfc6749.

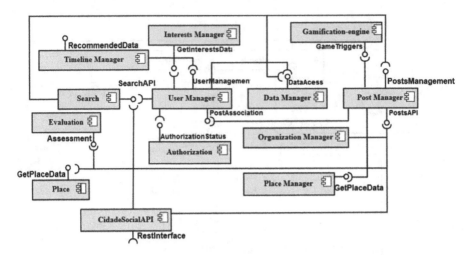

Fig. 4. Server component diagram.

all requests must be self-contained (they must have all information necessary for its processing).

The Timeline Manager component is responsible to set up the data which will form the user timeline. Each user has a personalized sample of a dataset according to their interests, the topic timestamp (the date and time of publication) and it is sorted considering the closeness between the user and the local where the topic was published (when GPS is available). If this process does not produce enough data, the client will receive the most recent data independent of the interest associated. This component depends on the features of the Data Manager and User Manager.

The Data Manager component provides access to services for the CRUD (create, update, read and delete) data. It receives the data from the clients (through Cidade-SocialAPI that uses the DataAccess interface) and registers the data in the server database. The access to the database is centralized in this component.

The resources related to the users are provided by the User Manager component, which is responsible for client authentication, new user registration and change of personal data (depending on the Authorization interface). This component is in fact an authorization credentials wrapper.

The Search component receives respective requests from clients (described above) and executes the query in the database. This component accesses the Data Manager component through the DataAccess interface.

The Evaluation component provides services to register client assessment in the database. It provides the Assessment interface. Similarly, the Place Manager and Organization Manager components provide the functionalities to deal with data concerning places and organizations unity.

The Post Manager registers and spreads the information (building up the users' timeline). Topics are associated with the user (through the UserManagement interface), the scores and experience gained are recorded in the system (through the GameTriggers

Table 2. Examples of URLs of CidadeSocial API.

Request method	Path	Return at success
GET	"/users"	A list of 20 first users
POST	"/users"	To create a new user
PUT	"/users/{user_id}"	Update the logged user
DELETE	"/users/{user_id}"	The id of deleted user
GET	"/users/{user_id}"	The user of a given id
GET	"/users/{user_id}/interests"	The interests of a user
DELETE	"/users/{user_id}/interests/ {interest_id}"	Just the 200 OK success message
GET	"/users/{user_id}/posts"	A list of 20 first posts of a user
POST	"/topic/"	The id of new published topic

interface. The component returns a response with the new user experience and punctuation to update the client's app data.

The Localization Manager provides a search data service of published data near to the current location of a user who is using that service. In addition, it provides the search service for published data in a region/topic of interest to a user who is using this service. The Interest Manager provides the necessary means to register, change or consult such interests.

Finally the Gamification component scores users according to the operations performed by them. This score is used to rank CidadeSocial users. For example, if a user publishes a topic and it is evaluated 100 times, that user should receive a number of points and some type of insignia.

4 Example of Use

Here the case of a potential user of CidadeSocial who wants to know the mobile application and access the web page (hosted in the server) is considered. This section aims to describe some cases and situations where this application may be used. Along with this description, interfaces of two client versions that have already been developed will also be presented.

Figure 5(a) shows the introduction page of CidadeSocial. On this page, users and administrators can access a brief introduction of the system. Users may also register using this page and as a result they will be able to access the platform from their mobiles. Others features which can be accessed through the web client are restricted to the administrators, who will be able to access the system information in full.

A topic Management interface for administrators is shown in Fig. 5(b). The center column shows topics listed according to interests. To do this, a filter (on the right of the page in the figure) was used. A topic may contain images, videos and the user geo coordinates related to the user localization at the moment it was published. It is mandatory to choose interests (the ones presented to the user during his/her registration) for the topics. Optionally a topic can be associated to a place (previously registered) or to organizational units (also with options previously registered).

As seen in the sample screens in Fig. 6, the mobile application is focused on ordinary users. Figure 6(a) is the screen for users to choose their interests. Figure 6(b) and (c) show the login interface and the user timeline, respectively. The lower icons on the screens in Fig. 5(a) and (c) are the components of the navigation menu. Through this menu the user can access all the features of the system. Figure 6(a) is the mobile client interests screen. Figure 6(c) exposes the user to all the information referring to his/her profile: score, favorites, ranking, awards (views related to gamification) and

(a)

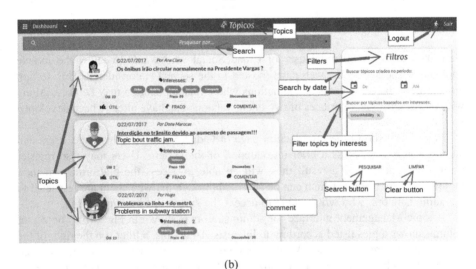

(b)

Fig. 5. Examples of pages from web client version.

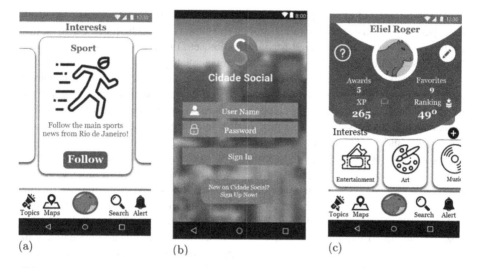

Fig. 6. Examples of mobile client screens.

mainly the interests that he/she selected when registering, which are used to generate the user timeline. The ranking is related to how engaged a user is and the scoring is related to the credibility of the user posts.

Considering the foregoing example of usage on a mobile device, we can consider a user who wants to track security-related topics on their timeline. After being authenticated on the system (an action that is handled by the User Management component) using the proper credentials, the mobile app will redirect the user to the home screen: Fig. 6(c). Then the user will select from the bottom menu the interests button and the system will return to the interests screen, shown in Fig. 6(a). On that screen, the full list of all interests available in the system is shown (Table 1). The user will view the list and select an interest and then, if the user is logged on, the security-related topics will be shown in the timeline. If there is no Internet connection, the synchronizer (Sync component) will take care of this as soon as it establishes an Internet connection (action controlled by the Connection Manager). The offline mode allows the user to use the system even without connection. The mobile application registers the actions performed and sends them to the server as soon as an Internet connection is identified.

5   Final Considerations and Future Works

This paper presented CidadeSocial, a tool for opportunistic and collaborative exchange of information using mobile devices. The published messages are related to interests from a predefined list in such a way that users can receive information according to their interests. This application is able to recommend information based upon the user's interests, search for information and encourage users in engagement using a gamification approach.

The differential of this proposal is its focus on interactions centered on user interests to support opportunistic collaboration among inhabitants that experience the dynamics (problems, changes, doubts, need for information, etc.) of cities. In this sense, CidadeSocial is an interface tool for community engagement [5].

The threats to the validity of this proposal are mainly related to widespread adoption and use. The use of the CidadeSocial in different contexts requires an intense user participation. Several barriers can prevent massive adoption. Some of them can be mitigated with respect to the usability of the interface, evaluations from users for adjustments and verification of intention to use.

As a future work we intend to conduct a study of the technology acceptance through a questionnaire with potential users with different profiles in order to reduce bias. Therefore, an evaluation questionnaire based on an experimental study plan applied to sample users will be elaborated to verify the perceived easy-of-use and usefulness of the CidadeSocial according the Technology Acceptance Model (TAM). Also we intend to evaluate the usability in order to improve user interaction with CidadeSocial. The development of client applications for the iOS platform is also foreseen, since native apps allow access to the device sensors. Finally, other improvements such as adding Labeling and Serendipity to recommend messages that are not part of the user interest group could be included in order to broaden the knowledge of his/her interests.

References

1. Agrawal, D., Budak, C., Abbadi, A.E.: Information diffusion in social networks: observing and influencing societal interests. In: Proceedings of the VLDB Endowment, vol. 4, no. 12, Seattle, Washington (2011)
2. Silva, T., Vaz De Melo, P.O.S., Almeida, J., Loureiro, A.F.: Large-scale study of city dynamics and urban social behavior using participatory sensing. IEEE Wirel. Commun. **21** (1), 42–51 (2014)
3. Mumford, L.: The City in History: Its Origins, Its Transformations, and Its Prospects. Harcourt Inc., San Diego, New York, London (1989)
4. Schaffers, H., Komninos, N., Pallot, M., Trousse, B., Nilsson, M., Oliveira, A.: Smart cities and the future internet: towards cooperation frameworks for open innovation. In: Domingue, J., et al. (eds.) FIA 2011. LNCS, vol. 6656, pp. 431–446. Springer, Heidelberg (2011). https://doi.org/10.1007/978-3-642-20898-0_31
5. Gordon, E., Mihailidis, P. (eds.): Civic Media: Technology, Design, Practice, pp. 563–580. The MIT Press, Cambridge (2016)
6. Yuan, J., Zheng, Y., Xie, X.: Discovering regions of different functions in a city using human mobility and POIs. In: Proceedings of the 18th ACM SIGKDD International Conference on Knowledge Discovery and Data Mining, pp. 186–194. ACM (2012)
7. Silva, T.H., Vaz de Melo, P.O.S., Almeida, J.M., Salles, J., Loureiro, A.A.F.: A comparison of Foursquare and Instagram to the study of city dynamics and urban social behavior, p. 1 (2013)
8. Kadar, C., Iria, J., Cvijikj, I.P.: Exploring Foursquare-derived features forcrime prediction in New York City. In: KDD - Urban Computing WS 2016, San Francisco, California, USA (2016)

9. Silva, E.R., França, T.C., Oliveira, J.: Aumento da Adesão e do Engajamento De Usuários do Campus Social Com Usode Mecanismos de Gamificação. In: Proceedings of the III Regional School on Information Systems of Rio de Janeiro, Seropédica, RJ (2016)
10. Deterding, S., Dixon, D., Khaled, R., Nacke, L.: From game design elements to gamefulness: defining 'gamification', p. 9 (2011)
11. Tabak, P., Figueiredo, E., França, T.C., Faria, F., Oliveira, J.: Campus social: uma ferramenta para trocas oportunísticas de informações em campi universitários. In: Anais do CSBC, Recife, PE (2015)

Structures of Interactions and Data in Urban Networks: The Case of PortoAlegre.cc

Pablo Vieira Florentino[1]([⊠]) and Gilberto Corso Pereira[2]([⊠])

[1] Federal Institute of Bahia, Campus Salvador, Brazil
pablovf@ifba.edu.br
[2] Federal University of Bahia, Salvador, Brazil
corso@ufba.br

Abstract. Urban spaces have been occupied by the massive use of new information and communication technology as digital social networks and platforms. The digital dimension of cities became a bidirectional and omnipresent path, creating relational and interactional structures able to exchange data and media. The networked city may be analyzed and debated as a complex system that demands research about communicational plurality and development of urban space representation, considering the increasing of informational and communicational density. Digital traces from social networks developed in PortoAlegre.cc, a collaborative web map registering issues and use of urban space, were used as data input for this research. Social network analysis was used as approach permitting network structures evaluation. The results reveal the existence of short paths, with predominance of structures that follow Small World model. The analyses showed efficient networks for data exchange increasing informational and communicational density. This work contributes for Urban computing bringing alternative approaches and perspectives for this multidisciplinary area with representation and knowledge that enhance the debate about urban space.

Keywords: Urban data fusion ·
Representation of collaborative social data ·
Volunteered geographic information · Digital social networks

1 Introduction

The City is certainly a complex study object, which can be associated to several meanings, specially if considered the many points of view from disciplinary fields composing what was used to call Urban Studies. These fields are usually occupied by architects, urbanists, geographers, sociologists, economists. Nowadays, Information Technology professionals have joined such fields developing several techniques and studies related to urban space and its data visualizations. The complexity of large contemporary cities may be characterized by an increasing diversity and amount of data which permit distinct modelling and representations. Technological convergence between media and communication made it available a large amount of digital data about the cities, produced by human activities – individuals or organizations – and automated systems. A presupposition of this paper is the acknowledgement of a large

© Springer Nature Switzerland AG 2019
J. Oliveira et al. (Eds.): BiDU 2018, CCIS 926, pp. 156–170, 2019.
https://doi.org/10.1007/978-3-030-11238-7_10

set of data produced in urban space by devices as cell phones, sensors (temperature, pollution, noise, GPS, etc.), social networks (Facebook, Twitter, Instagram, etc.) and others used to generate geolocated information. It's relevant to emphasize the central role of geographic localization in a context of dynamic appropriation of data provided by connectivity infrastructure.

Physical structure and infrastructure, as public spaces, squares, streets, private spaces, mobility, density, segregation, land use regulation keep relevant to Urban planning and management activities. Nevertheless, it is necessary to add alternative tools, able to show and manage the contemporary context of information flows, connecting businesses, governmental and non-governmental organizations and, especially, citizens, whom, from these cultural and technological possibilities, can also act politically and influence trends and actions on these structures. In face to such demands, management and planning of urban space shall involve, nowadays, sophisticated tools, processes and, specially, knowledge derived from Big Data, leading to knowledge fusion [10] supported by urban computing joined to a data science area [6].

In this context of imbrications and overlapping between structures and dynamics, structures of communication and socialization into cities became an issue to observe and analyse, considering the digital tiers of urban space influence on the enlargement of contexts and fluxes of information. During the investigation process, we have perceived the absence of representations able to capture social and communicational relations developed in such digital tiers and to analyse their own structures.

Ubiquity of connections and technological convergence are promoting a set of sociocultural practices adding new meanings to information with the amplification of production of what we call *information in space*. In digital culture, distance can be measured in different ways. Access to digital spaces where transactions occur, social and political interact, academic and cultural activities are measured in "clicks" or "at the touch of a button" [5]. The analysis of networks may guarantee other ways of observing such distances between elements of networks established in the digital sociability. The implications of these changes are both cultural and technological, restoring McLuhan's old argument: "the medium is the message" [13], and must be considered in current urban planning practices. By allowing user interaction with physical environment representations and wide information exchange between them, information and communication technologies facilitate the involvement of individuals in participatory activities through the Internet. For instance, collaborative mapping and crowdsourcing, location based social networks and geographic tools on mobile devices, augmented reality applications form a list of a few possibilities and challenges [10, 18].

Based on such findings and assuming Internet as a technological infrastructure already naturalized, therefore, a contemporary urban culture element, we examine in this work ways of representing city knowledge by its own inhabitants, considering the city of Porto Alegre (capital of the State of Rio Grande do Sul/Brazil). The case study on users communities of collaborative mapping aimed to identify specific relationships among locations (neighbourhoods), and citizens contributing to the construction of such digital cartographies. We used a collaborative map of Porto Alegre, **PortoAlegre. cc**, a cartographic, collaborative and voluntary application, open to visitors. Any individual – registered on the site – could create georeferenced causes on urban space of Porto Alegre through its interface (Fig. 1).

Thus, our main objective in this work is analysing how social networks formed in digital spaces, like PortoAlegre.cc, organize the relational space of cities and amplify the contexts and densities of relations among elements in urban spaces, observing their usages, fluxes, traces and structures for generating alternative representation and perspectives about the city. We follow the hypotheses of the geographer Milton Santos, who states that informational and communicational density becomes amplified [22], considering the structures arising in digital networks. Our secondary objective is focused on systematizing and executing such a proceeding that enables, with more methodological rigour, arranging and confronting quotidian impressions versus certain spatial urban practices.

2 Urban Networks and Social Data

The emergent civic structures and spatial arrangements of digital age affect access to economic opportunities and public services, the bias of public discourse, cultural activity forms, institution of political power and experiences shaping and filling the content to our daily routines [14]. Even city representations may cover a broad range of aspects, i.e. as physical aspects – city's topography, in a broad sense – or social ones – social demographic representations, social networks, personnel or institutional interactions, Sassen observes that topographical representations, that is, representations of physical reality, fail to capture the essence of contemporary economic, political and cultural relations [23]. Traditional representations of social and demographic aspects in cities are based on aggregation of individuals or families into defined and uniformly represented areas (census tracts, for example) or the physical address of individuals such as those in commercial databases. While this form of data and information processing remains relevant, contemporary digital social networks (DSN) are playing a structural role in society and producing larger, diverse and enriched volumes of social data coming from urban networks. Social interactions from everyday life are becoming more digital technologies dependent, which are reinforced by DSN and contribute to organize daily life and sociability. It makes almost mandatory the need for connectivity and portable devices to access the distributed interfaces in urban infrastructure. This presents new challenges for the construction of spatial representations of society and its relations, with implications in activities that work with geodemographic data [25], as urban planning and management.

It is the role of urbanism to realize the possibilities that arise with such contemporary and technological context in which the city has its space almost totally digitized and may be visualized in diverse forms of representation. Examples of the current uses of spatial representation forms indicate a possible change in the perception of contexts by social sharing and by the increasing dynamic integration of information. The representation of space and the represented space come closer, overlapped and intercepted. Models, cartographies, georeferenced images, videos, tweets and posts coexist in urban space with the structure and physical infrastructure. The issues related to the representation and knowledge of space can be perceived in a cycle that leads to the representation of knowledge through the use of technology and this, in turn, leads to the enrichment of the context and knowledge of the (urban) place. DSN are now the space where people, especially young people, connect, communicate, exhibit and interact much more than in the streets, squares or shopping centres [7, 17].

In such urban context, cities may be understood as complex systems [8, 19, 20, 27] crowded of several emergent processes structured in different kinds of networks. Thus, we have chosen a network analysis based approach in order to develop the present study, reinforcing the multidisciplinary nature of this work, making use of mathematics and computational techniques for modelling and representation of urban relational space dynamics. Such approach offers a distinct formulation of traditional analysis and representation techniques of urban space and its social data, mainly considering relations among actors and urban elements with communicative exchanges. It represents a feasible way for achieving the defined goals, utilizing digital traces from a specific DSN (PortoAlegre.cc) to expose its network structure and characteristics. This would permit classification and more detailed interpretation of social behaviour and organization in such citizens community and their relations with neighbourhoods from Porto Alegre city.

3   PortoAlegre.cc

PortoAlegre.cc, a virtual map focused on urban space and representing the city of Porto Alegre, was launched as an open and cartographic web application (Fig. 1), permitting any person registered in the website to create georeferenced causes about urban space in that city. A mural of occurrences and public events in Porto Alegre was composed by contributions and interactions related to positive or negative facts and characteristics of distinct aspects (the so called *causes*).

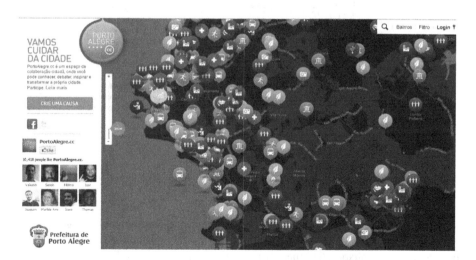

Fig. 1. Interface of collaborative project PortoAlegre.cc

This digital georeferenced mural could provide input for public administration, so far, making possible digital links among citizens without direct relationships in off-line life, trough interactions and information exchange via the cartographic web application. This process also permitted articulation and participation in collective activities of intervention and occupation in urban public sights, fostering debates about urban issues [24]. The platform has raised up social participation, promoted collaborative spaces by behaviour exposure of the city in a categorized and georeferenced form, aggregating new visions of urban space.

Each cause was classified in one of the available themes related to urban life: Citizenship, Culture, Education, Entrepreneurship, Sports, Environment, Mobility, Health, Safety, Technology, Tourism and Urbanism, distinguished by colour, as presented in map interface (Fig. 1). The causes, described by texts and pictures, registered different issues such as thefts, infrastructure problems in public streets, non-functional public equipments, notification and organization of sportive or leisure meetings or events for collaborative recuperation of non-assisted areas by public administration. The full content of the digital map involves 932 participants and its causes descriptions were visible to anyone – even non-registered citizens – and each cause had, on average, 530.17 visualizations and 5.73 *"likes"* from non-registered visitants. The distribution of causes in the different themes is presented in Fig. 2, where the themes of Mobility, Citizenship, Environment and Public Safety head the most cited aspects.

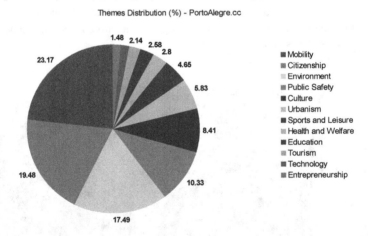

Fig. 2. Percent distribution of causes by themes (Color figure online)

This digital platform may be classified as a decentralized volunteered geographic information system (VGI) [4], free of restrictions to registration or what shall be registered on the collaborative map. It is also categorized as a feedback system for the society and public administration. The data captured from PortoAlegre.cc represents a relevant part of urban reality and dynamics in Porto Alegre city, enabling data structuring and characterization in imperceptible levels, until so, considering there was no open mechanisms of wild range in time-space in such period. The testimonials from the creators of the collaborative map [24] reported lots of volunteers in person meetings after digital interactions and articulations. Moreover, unknown volunteers established social ties based on citizenship practice, allowing social exchanges that would hardly happen without such a collaborative platform.

The whole collaborative and participative process developed using the PortoAlegre.cc platform involves hundreds of distinct social actors who, together, produced a collective intelligence about the city generated from its own urban space. Such process mapped and shared issues and aspects that, for most part, do not belong to the main interests and concerns of public administration (public safety and mobility, i.e.). PortoAlegre.cc becomes, thus, an element for public disclosure and diffusion of such issues, increasing and detailing the urban space dynamics, now represented in a digital dimension and broadly shared.

4   Methodology

We have used a database provided for academic purposes and concerning hundreds of collaborations into the digital map considering a period of 18 months, involving 932 participants. In such sample, a first procedure was made for extracting neighbourhood from textual address of each cause. Considering each neighbourhood could have several causes, there were 79 neighbourhoods from Porto Alegre identified in all causes textual localization. A second procedure retrieved latitude and longitude from each neighbourhood.

Our approach was based on Social Networks Analysis (SNA) methods [1, 2, 11, 16, 26] for structuring and analysing the collected data considering localization, themes, volunteers causes, in a way to model relationships among: (a) volunteers; (b) neighbourhoods and volunteers; (c) neighbourhoods. Our assumption is that such relationships were, until then, imperceptible and/or not susceptible to modellings by traditional methods. Furthermore, considering data from Porto Alegre City Observatory[1], the respective codes from each Participatory Budget Region and each Planning Region were added to original data, aiming to observe how distinct regions are linked by volunteered information. The distinct perspective, adopting city organization in neighbourhoods and regions mixed with voluntary data registered in the collaborative map, is a challenge issue to identify possible indirect relations among inhabitants in urban space. We have generated a two-mode (bipartite) network [11]. A bipartite network is a graph $G = (\top, \perp, E)$ where $\top$ is the set of *top* nodes, $\perp$ is the set of *bottom* nodes, and $E \subseteq \top \times \perp$ is the set of links. The difference with *classical* networks lies in the fact that the nodes are in two disjoint sets, and that the links always are between a node of one set and a node of the other. In other words, there cannot be any link between two nodes in the same set. In our work, for each cause created by a volunteer from PortoAlegre.cc in a determined neighbourhood from Porto Alegre city, an edge was created linking the volunteer to the referenced neighbourhood. We have

---

[1] Data available at www.observapoa.com.br.

also spatialized the cited neighbourhoods in comparison to the data distribution of socio-spatial typology of dwelling place, according to the classification presented in [12], in the sense of obtaining the socio-spatial profile of the collaborative map. From the original 2-mode network, we have generated the projection of two networks of 1-mode: neighbourhoods network and volunteers in the collaborative platform network. Such networks will contribute for identifying: (1) how volunteers are connected by localities in the city; (2) the non-trivial relationships among neighbourhoods.

We aim to evidence the formation of networks based on georeferenced narratives from Porto Alegre inhabitants – users of the volunteered platform – characterized as indirect networks among their participant elements. Thus, the structure of such collaborative and voluntary networks were represented by models able to provide information about urban space dynamics and inferences about how people are indirectly connected through their geolocated participation. The data spatialization indicates knowledge about the most cited places and how the dynamics of spatial moving in Porto Alegre occurs.

Considering the social practices collected through digital traces from PortoAlegre.cc, a neighbourhood oriented network of volunteers was modelled and evidenced unknown inhabitants that keep indirect links. Such modellings organize narratives of individual comprehension about the urban space and the geolocated citations arranged in a network, enabling to forge strategies of citizens traces and displacement visualization. From the preliminary steps of research and their results, we have perceived the opportunity and need for broadening our studies. For improving the portraying of such spatial arrangement usage, we developed a heat map corresponding to geographic distribution of causes registered in PortoAlegre.cc.

5   Results

In the 2-mode network generated (Fig. 3), volunteers just establish relations with neighbourhoods and vice-versa. The vertex sizes are proportional to respective degrees of relationships. The neighbourhoods (in yellow) are highlighted, concentrating most part of relationships with volunteers from PortoAlegre.cc.

Such connections agglomeration around neighbourhoods is justified by the main motivator and major interactional via: the collaborative map representing the own urban space. A few users (in blue) have some prominence in the network (whose names are omitted for confidentiality), demonstrating more activity in the map.

The neighbourhoods with higher degree centrality are also neighbourhoods localized in regions of higher urban centrality of Porto Alegre city: Centro, Praia de Belas, Petrópolis, Farroupilha, Menino Deus, Rio Branco, Bomfim, Cidade Baixa, Partenon and Arquipélago [21]. The same effects and standards of centrality, characteristic of larger urban centres, are repeated in the usage of PortoAlegre.cc, revealed by the heatmap in Fig. 4.

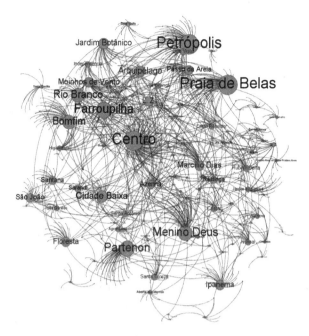

Fig. 3. 2-mode network: volunteers (PortoAlegre.cc) X neighbourhoods from Porto Alegre (Color figure online)

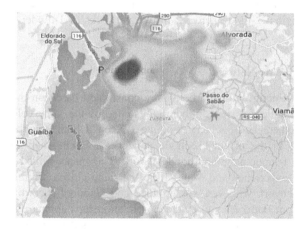

Fig. 4. Heat-map of causes created in PortoAlegre.cc

The red areas highlight the regions with higher concentration of causes, while the green ones evidence those with lower number of causes. The comparison to distribution of socio-spatial typology, according to [12], also disclosed a concentration of causes around areas of higher standards of dwelling (Fig. 5).

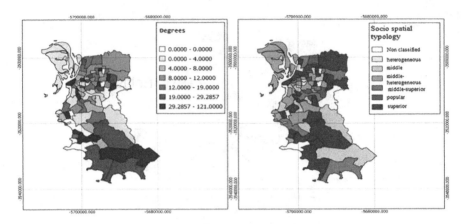

Fig. 5. Porto Alegre, neighbourhoods: degree of relationships X socio-spatial typology (Color figure online)

Even neighbourhoods in central regions have gathered most part of causes, a set of factors indicate an amplification of contexts and social relationships, increasing informational and communicational density, as stated in [22]. We consider the high number of visualizations and interactions, the occurrence of causes in almost all neighbourhoods from Porto Alegre and spatial representation in a cartographic inter-face, allowing amplified vision of the city and its dynamics. Nevertheless, we under-stand the need for a more detailed investigation on indirect relations, evidenced by projections of original 2-mode network in two networks of 1-mode: neighbourhoods and volunteers.

5.1   Neighbourhoods Network

The relationships among neighbourhoods are constituted based on the links between volunteers and localities: those neighbourhoods cited by the same volunteer define an edge with each other. Thus, when a volunteer X cited both neighbourhood A and neighbourhood B in the original 2-mode network, a linking between A and B is established in the 1-mode neighbourhood network. This means linked neighbourhoods are part of urban experience and routes from the same citizen. Figure 6 presents the neighbourhood network with vertexes grouped and coloured according to municipal budget regions from Porto Alegre, evidencing relationships among distinct regions.

The average degree of relations in this network shows that one neighbourhood has been cited jointly with 22 other neighbourhoods, demonstrating that volunteers dis-tributed their contributions on distinguished neighbourhoods, pointing out to a diver-sified displacement behaviour in the city. Figure 7 brings a spatial representation of neighbourhoods network, overlapped to neighbourhoods map from Porto Alegre.

Following the standard procedures for SNA, the results of minimum average path and average clustering coefficient and the behavior of degree distribution permitted us to classify such network as a Small World Network. This permits some interpretations of such network structure: (a) subgroups of neighbourhoods in which happens a higher

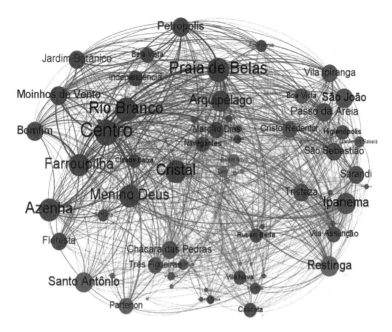

Fig. 6. Neighbourhoods network (projection from original 2-mode network) (Color figure online)

flow of the same volunteers; (b) elements in such network functioning as bridges among subgroups, enabling smaller distances among the other elements; (c) efficient paths from volunteers among the cited neighbourhoods in the collaborative map.

Considering the total of links, 87.16% of them are among distinguished budget regions and 82.44% of them are among distinct municipal planning regions. It's also perceptible a large number of neighbourhoods in geographic remote regions, but linked by locative practices registered by volunteers, generating digital traces in different regions from Porto Alegre. In other words, neighbourhoods pairs cited by, at least, one common volunteer and localized in different regions represent the larger part of links among neighbourhoods.

The aspects disclosed by the network structure (Fig. 6) and by the same network superimposed to the city's map (Fig. 7) reinforce the locational diversity of activities over PortoAlegre.cc, indicating a higher plurality in the urban space represented and shared in such an open platform. This demonstrates a greater coverage of PortoAlegre. cc, going beyond urban centralities from Porto Alegre, reaching disperse and distinct areas.

The betweenness index – frequency with which a node falls between pairs of other nodes on the shortest paths connecting them [26] - disclosed other neighbourhoods in highlighted positions (Fig. 8), as Lageado. Even without a high number of links (few citations on map), it owns a high betweenness value. This neighbourhood becomes an important linking element in locational practices from volunteers and neighbourhoods subgroups. Its removal from the network would disconnect several subgroups and regions or would increase distances among them.

Fig. 7. Neighbourhood network overlaid on Porto Alegre map

5.2   Volunteers Network

The 1-mode volunteers network was projected and represented in Fig. 9, on which each link between two volunteers means both collaborators have created at least one cause over the same neighbourhood. Such visualization discloses indirect links volunteers and their intersections of space usage. Aiming an improved arrangement, we have executed a procedure for clusters identification, based on the Chinese Whispers algorithm [3]. 30 clusters were identified and represented in different colours, reflecting some concentrations in locational practices on the map, reinforcing a high average clustering index.

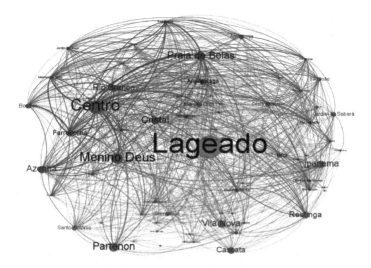

Fig. 8. Neighbourhoods network visualization considering betweenness (Color figure online)

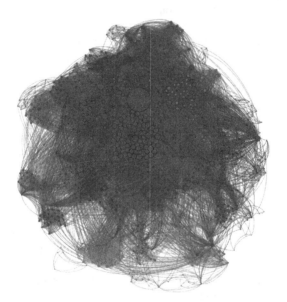

Fig. 9. Volunteers networks from PortoAlegre.cc (Color figure online)

The average degree of links shows that, on average, a volunteer from PortoAlegre. cc created causes in the same neighbourhoods 49 other volunteers did, allowing interactions and knowledge exchange about cited locations. The short paths in such urban network – among those who collaborate with the map – increased the probability of data information exchange or creation of some kind of connection beyond the collaborative platform. Thus, volunteers had more chances of closeness into urban environment via interactions exchange in digital dimension of the city.

The values of average clustering coefficient and average shortest paths indicate a network structure that also follows Small World model. This permits us to advance in some issues: subgroups of volunteers citing more neighbourhoods in common; volunteers as bridges among subgroups, making distances shorter among networks elements; higher efficiency on information exchange.

6   Considerations

The urban space is often apprehended as an open and given element, over which are settled buildings and people. We comprehend the spatial issue as a more complicated problem: observing the interference of a georeferenced data layer, our work presents the space crossing (more then what it is crossed by) distinguished citizens, performing the role of a social chain among such volunteers. Besides being a daily visiting, in passing or meeting point, a certain neighbourhood interlinks different actors, serving as bridge or sociocultural axis. The experience context in the city is amplified for a higher detailing level considering collective visions and opinions, going beyond traditional media or daily living and percusses from volunteers. Such dynamic may be, obviously, perfectly deductible in face of neighbourhoods with larger human clusters. What we did, effectively, was modelling indirect relationships among social actors (even they did not have previous social relations). The network oriented approach enabled the perception of relationships conducted by informational mediation of each location. Such urban networks disclosed favourable structures to the increasing of informational and communicational flows, with short paths among their elements, even if they are located or inhabit geographically distinguished regions. This confirms the amplification of contexts theory presented in [22].

We have verified, from such networks, the possibility of identifying and investigating the profiles that approximate by locative practices in urban space and exploring their common characteristics as gender, regions of dwelling, topics of interest, among others (a possible future work for this research). For instance, the major part of causes on the map are situated in regions with socio-spatial typology of higher standard [12], revealing tendency to a higher socioeconomic profile of volunteers. It's also possible to generate and spatialize thematic networks and observe their structures, helping public administration to identify gaps in urban planning and improving their knowledge about citizen profile.

By considering this case a community of collaborative mapping, we have also verified the existence of volunteers subgroups that registered causes in specific neighbourhoods groups from Porto Alegre. Some elements in such network cited a higher number of neighbourhoods in distinct regions, becoming bridges among subgroups with a diversified behaviour trough the map. Small World Networks, like those from PortoAlegre.cc, with topics of interest driven to debate the urban space or promoting different ways of utilizing the city, own a relevant position in developing such a discussion. This kind of network is more efficient in information transmission, according to SNA theory, improving the expansion of social dialogues and knowledge fusion representation about the city, besides allowing more social interaction, influencing the increasing of urban informational and communicational density.

We also noticed relevant social actors functioning as interactions bridges with a spread behaviour in distinct neighbourhoods from different regions, revealing how some neighbourhoods, with few citations on PortoAlegre.cc, function as mediators among regions. That's the case of Lageado, Fig. 8, that, even with few citations, is part of many possible paths linking other neighbourhoods, performing a highlight role in betweenness.

The contemporary practice of socializing perceptions about the city has been converted into more broadening changes in forms of comprehending the urban space, as well as influencing the occupation of spaces into the city. The interchanges promoted by PortoAlegre.cc allowed connections among unknown inhabitants, an impossible fact without the growth and consolidation of digital platforms and DSN as communicative channels. When interpreted trough network oriented approaches and techniques, it was possible to understand the collective behaviour in such informational and interactional spaces, which generate non premeditated changes in a scale above the individual or subgroups level, as state Mitchell [15] and Johnson [9] This fact reveals an emergent behaviour, confirming such case study as a complex system. Once modelled and characterized as a network, a system of people who join, without a central control, generating complexity on collective behaviour, intrinsically tied to the relationships among its elements, and not only to the particular social role each participant performs. The disclosure of urban issues, generating inquiries from society related to public administration effectiveness, shall also be considered as a result from an emergent and collective behaviour and an input for urban planning.

The procedure of the present work exposes geographic data as an agglutination and correlations element among several citizens and shows, partially, how locations and spatial practices are reaching a higher level of complexity, amplified and expanded with usage of electronic technologies and generating a large amount of data. Whether localities were, before, merely places to reach, they passed to constitute informational layers relating and merging several actors and elements generators of urban data. We highlight, thus, the relevance of urban computing processes on gathering and analysing social and geographic information as an element for social clustering and potential producer of sociability. It is a relevant kind of data in contemporary digital culture passing transversally other kinds of data – in other words, becoming incorporated as metadata on several contents from urban space. This work presented non-trivial multidisciplinary studies about urban spaces and its dynamics, involving urban computing and social data. Urban planning and public administration can benefit from such distinct perspectives, opening possibilities to improve and guide decision making processes.

References

1. Barabasi, A.L., et al.: Human mobility, social ties, and link prediction. In: ACM SIGKDD (2011)
2. Barabási, A.L., Bonabeau, E.: Scale-free networks (2003)
3. Biemann, C.: Chinese whispers - an efficient graph clustering algorithm and its application to natural language processing problems. In: Proceedings of the First Workshop on Graph Based Methods for Natural Language Processing, pp. 73–80. Association for Computational Linguistics, Stradsburg (2006)

4. Craig, W.J., et al. (eds.): Community Participation and Geographic Information Systems. Taylor & Francis, London (2002)
5. Devriendt, L., et al.: Cyberplace and cyberspace: two approaches to analyzing digital intercity linkages (2008)
6. EMC Corporation (ed.): Data science and Big Data Analytics: Discovering, Analyzing, Visualizing and Presenting Data. Wiley, Indianapolis (2015)
7. Florentino, P.V., et al.: City as a social network – Brazilian examples. In: UDMS 2013 - 29th Urban Data Management Symposium. CRC Press/Balkema, London (2013)
8. Jacobs, J.: Morte e vida de grandes cidades. Martins Fontes, São Paulo (2000)
9. Johnson, S.: Emergência: a vida integrada de formigas, cérebros, cidades e softwares. Zahar, Rio de Janeiro (RJ) (2003)
10. Kamienski, C., et al.: Computação Urbana: Tecnologias e Aplicações para Cidades Inteligentes. In: XXXIV Simpósio Brasileiro de Redes de Computadores e Sistemas Distribuídos, pp. 51–100. Sociedade Brasileira de Computação, Salvador (2016)
11. Latapy, M., et al.: Basic notions for the analysis of large two-mode networks. Soc. Netw. 30 (1), 31–48 (2008)
12. Mammarella, R., et al.: Estrutura Social e Organização Social do Território: Região Metropolitana de Porto Alegre – 1980–2010. In: Fedozzi, L., Soares, P.R.R. (eds.) Porto Alegre: transformações na ordem urbana, pp. 133–184. Letra Capital Editora, Rio de Janeiro (2015)
13. McLuhan, M.: Understanding Media: The Extensions of Man. MIT Press, Cambridge (1995)
14. Mitchell, W.J.: City of Bits: Space, Place, and the Infobahn. MIT Press, Cambridge (1996)
15. Mitchell, W.J.: E-topia - A Vida Urbana - Mas Não Como a Conhecemos. Senac (2002)
16. Newman, M.E.J.: Networks: An Introduction. Oxford University Press, Oxford (2010)
17. Pereira, G.C., et al.: Accessing the city through new forms of sociability - examples of use of digital social networks in Brasil. Territorio Italia 2, 71–83 (2015)
18. Pereira, G.C., Rocha, M.C.F.: Spatial representations and urban planning. In: Planning Support Tools: Policy Analysis, Implementation and Evaluation, Roma, pp. 611–623 (2012)
19. Portugali, J., et al.: Complexity Theories of Cities Have Come of Age. Springer, Heidelberg (2012). https://doi.org/10.1007/978-3-642-24544-2
20. Portugali, J.: What makes cities complex? In: Complexity, Cognition, Urban Planning and Design. TU Delft, Delft (2013)
21. Prefeitura de Porto Alegre: História dos Bairros de Porto Alegre. http://lproweb.procempa.com.br/pmpa/prefpoa/observatorio/usu_doc/historia_dos_bairros_de_porto_alegre.pdf (2006)
22. Santos, M.: A natureza do espaço: técnica e tempo, razão e emoção. Editora da Universidade de São Paulo, São Paulo (1996)
23. Sassen, S.: The city: between topographic representation and spatialized power projects. Art J. 60(2), 12–20 (2001)
24. Secco, D.: Entrevista presencial sobre o projeto PortoAlegre.cc (2013)
25. Singleton, A.D., Longley, P.A.: Geodemographics, visualisation, and social networks in applied geography. Appl. Geogr. 29(3), 289–298 (2009)
26. Wasserman, S.: Social Network Analysis: Methods and Applications. Cambridge University Press, Cambridge (1994)
27. Modeling Cities and Regions as Complex Systems. The MIT Press. https://mitpress.mit.edu/modeling-cities

DMEK: Improving Profile Matching in Opportunistic Collaborations

José Guilherme Mayworm$^{(\boxtimes)}$, Jonice Oliveira, Fabrício Firmino,
and Claudio M. de Farias

Universidade Federal do Rio de Janeiro, Rio de Janeiro, RJ, Brazil
jgmayworm@gmail.com, jonice@gmail.com, firminodefaria@gmail.com,
claudiofarias@nce.ufrj.br

Abstract. As the number of mobile devices grow, also grows the amount of data exchanged. This ever growing amount of data may overload Internet Service Providers. A possible solution to this problem is to use the mobile devices wireless network capabilities to exchange data by creating mobile P2P networks. These networks should opportunistically collaborate to exchange information to other devices in their proximity, only requiring users to specify their interests. This paper presents DMEK, (Decision Mobile Exchange of Knowledge) a solution where mobile devices disseminate knowledge among their users, opportunistically, using a decision mechanism based on profile matching. Experiments show DMEK feasibility and performance.

Keywords: Opportunistic · Collaboration · Mobile devices ·
Profile matching · Peer-to-peer · Exchange · Knowledge

1 Introduction

In 2016, the number of smartphones sold to consumers stood at over 2.1 billion units, as shown in Fig. 1, an increase of over forty percent from 2014 sales. This means that almost sixty percent of the world's total population owned a smartphone in 2016 [1]. The integration of a variety of sensors (motion sensors, accelerometers and gyroscopes) and wireless short-range technologies (such as Bluetooth, NFC, RFID and 802.11) allowed these portable devices to stay permanently connected to the internet, and sense physical environments without human direct interaction. These characteristics made them perfect candidates to work in the *Internet of Things* (IoT) [2]. The concept of IoT refers to enhancing physical objects with computing, sensing, and communication capabilities, into smart objects and connecting them to form a network and through their collective collaborative effort, reach common goals [3].

Traditional IoT attempts to connect smart objects by using pre-existing infrastructure. Alternatively, these devices can be connected by using in-built wireless communication to form decentralized, mobile adhoc networks [4].

© Springer Nature Switzerland AG 2019
J. Oliveira et al. (Eds.): BiDU 2018, CCIS 926, pp. 171–184, 2019.
https://doi.org/10.1007/978-3-030-11238-7_11

In this sense, smartphones present an opportunity to build these networks as they are commercially available and are coupled with required hardware. Leveraged by proper software, the devices can autonomously communicate to form mobile P2P networks [5,6], and opportunistically collaborate to exchange information to other devices in their proximity, only requiring users to specify their interests. In the context of this work, we define opportunistic collaborations as people engaging together in spontaneous and unexpected interactions, that only happen when the participants meet, to achieve common goals.

In order to achieve this collaboration some solutions were proposed [7]. Among these solutions we highlight the Mobile Exchange of Knowledge (MEK), a solution where smartphones engage in short-timed opportunistic exchanges of data, based on the interests of their users [8]. MEK autonomously share pieces of knowledge (user generated content such as images, text, or audio) as people commute about in their daily routines. The application aims to increase the exchange of knowledge among peers, by forming mobile P2P networks composed of users sharing the same interests. These networks, however, are highly dynamic because the nodes move arbitrarily, causing the network topology to change frequently and unpredictably. As a consequence, discovering resources becomes challenging since connections among peers are short-lived, giving little time for applications to retrieve information efficiently [5].

For instance, to exchange knowledge MEK compares profiles by keyword matching interests users assigned to their content. This approach is inefficient because any match (without any consideration about semantic compatibility) will trigger exchanges, even if they share few interests. If the network grows, the number of connections and data transfers will also increase, overloading both the network and the devices.

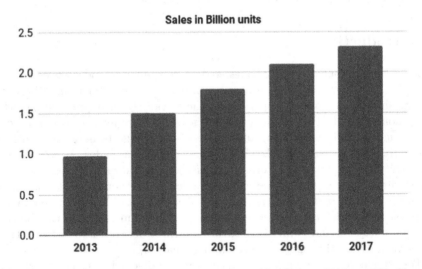

Fig. 1. Number of smartphones sold to end users worldwide from 2013 to 2017. (Source: www.statista.com)

Therefore, in this work, our goal is to improve the decision-making process of MEK to perform exchanges based on profile similarity. To achieve this we use a data fusion technique, called Dempster-Shafer, to combine information from profiles in order to determine their similarity. Dempster-Shafer was also used due to its low computational cost and faster answer [2], which are really important in the aforementioned scenario. We propose an extension of MEK, hereinafter called DMEK (Decision MEK) that uses a decision system to determine a good peer to trade information. We will classify the peers based on its similarity score.

This work is structured as follows: Sect. 2 we present the basic functioning of MEK and the problem of profile matching with keywords in further detail. In Sect. 3 we present some related works. In Sect. 4 we propose our approach towards profile matching with data fusion. Following in Sect. 5 we describe implementation details, evaluation metrics and results. Finally in Sect. 6 we present our conclusions and future works.

2   Basic Concepts

In the following sections we will present the definitions and backgrounds of concepts in this work.

2.1   The Mobile Exchange of Knowledge

MEK is a application in which mobile devices are used to disseminate knowledge among its users opportunistically, according to patterns of movement and encounter of human beings. The purpose of the application is to exchange user generated content such as, images, text, or audio in a proactive and viral way. MEK aims to increase the exchange of knowledge, forming a mobile peer-to-peer network composed of users sharing the same interests. Once the application is installed, users have to identify their interests by filling in a small form, indicating their areas of interest, relevant keywords, and other information that will be used to build their personal profile. The interests are represented by concepts in a preset taxonomy where an interest may contain several sub-areas of interest. After building the user profile, users could subscribe to knowledge topics delivered to their device. This knowledge will be matched according to the taxonomy and the keywords from the user, to help improve classification. The basic functioning occur as follows: to exchange information, a MEK device periodically scans for others in the vicinity running the application. When another device is found, the profiles of local users and their interests are exchanged. If there are any matches, the selected knowledge is sent to the requester. Figure 2 describes how MEK identifies potential candidates for exchanges.

MEK uses Bluetooth to establish communication. The motivation to use this technology, as Oliveira says, is because it is widespread [9]. It also allows for flexibility, and since it works in lower transmission ranges and data rates than 802.11, it has lower cost and lower power consumption. On newest versions, Bluetooth has significantly increased range, speed and broadcast messaging capacities [10].

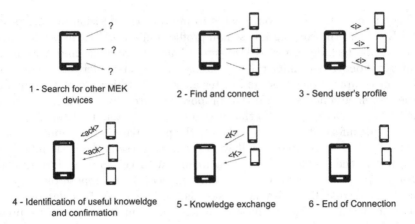

Fig. 2. How knowledge is exchanged in MEK, where <i> is information about user's profile, <ack> is acknowledge and <k> is a piece of knowledge.

Next, we describe in further detail the problem of exchanges among users who share few similarity.

2.2 The Profile Matching Problem

DMEK's main goal is to autonomously share information, as people move around in physical locations. Naturally the application can establish a high number of connections, specially in scenarios where the concentration of people is high (shopping malls, airports, stadiums, music concerts, among others). The issue with this approach is that any devices that enter in range of communication and share at least one interest will prompt transmission of data. In this sense, the application performs many exchanges while disregarding the similarity of profiles as a whole. As a consequence, a great work-load is imposed on devices without obtaining comparable results. Even though smartphones have become more powerful recently [11], this approach might still negatively impact their resources, such as draining battery and overloading CPU. One way to mitigate this problem is to improve the decision process of MEK towards evaluating profile similarity. This way we can filter relevant profiles from the crowd and diminish the number of exchanges. Since MEK is envisioned for dynamic scenarios, finding relevant profiles must be timely, in order to detect devices in movement, and efficient in determining users similarities. In the next section we present a data fusion technique called Dempster-Shafer.

2.3 Dempster-Shafer

Dempster-Shafer Belief Accumulation (also referred to as Theory of Evidence or Dempster-Shafer Evidential Reasoning), is a general framework for reasoning with uncertainty, that generalizes the Bayesian theory [12,13]. It deals with beliefs or mass functions just as Bayes' rule does with probabilities. The Dempster-Shafer theory provides a formalism that can be used for incomplete knowledge representation, belief updates, and evidence combination [14]. A fundamental concept in Dempster-Shafer reasoning system is the frame of discernment, which is defined as follows. Let $\Theta = \{\theta_1, \theta_2, \ldots, \theta_N\}$ e the set of all possible states that describe the system, such that Θ is exhaustive and mutually exclusive in the sense that the system is certainly in one, and only one, state $\theta_i \leq \Theta$, where $1 \leq i \leq N$. We call Θ the frame of discernment because its elements are used to discern the actual system states.

The elements of the power set 2^Θ are called hypotheses. In the Dempster-Shafer theory, based on evidence E, a probability is assigned to every hypothesis $H \in 2^\Theta$ according to a basic probability assignment (BPA), or mass function, $m : 2^\Theta \to [0, 1]$ that satisfies:

$$m(\emptyset) = 0 \tag{1}$$

$$m(H) \geq 0, \forall H \in 2^\Theta \tag{2}$$

$$\sum_{H \in 2^\Theta} m(H) = 1 \tag{3}$$

To express the overall belief in a hypothesis H, Dempster-Shafer defines the belief function $bel : 2^\Theta \to [0, 1]$ over Θ as:

$$bel(H) = \sum_{A \subseteq H} m(A), \tag{4}$$

where $bel(\emptyset) = 0$, and $bel(\Theta) = 1$. The degree of doubt in H can be intuitively expressed in terms of the belief function $bel : 2^\Theta \to [0, 1]$ as:

$$dou(H) = bel(\neg H) = \sum_{A \subseteq \neg H} m(A). \tag{5}$$

To express the plausibility of each hypothesis, the function $pl : 2^\Theta \to [0, 1]$ over Θ is defined as:

$$pl(H) = 1 - dou(H) = \sum_{A \cap H = \emptyset} m(A) \tag{6}$$

The plausibility intuitively states that the less the doubt in hypothesis H, the more plausible. In this context, the confidence interval $[bel(H), pl(H)]$ defines the true belief of the hypothesis H. To combine the effects of two bpa's m_1 and m_2, the Dempster-Shafer theory defines a combination rule, $m_1 \oplus m_2$, which is given by:

$$m_1 \oplus m_2(\emptyset) = 0, \tag{7}$$

$$m_1 \oplus m_2(H) = \frac{\sum_{X \cap Y = H} m_1(X) m_2(Y)}{1 - \sum_{X \cap Y = \emptyset} m_1(X) m_2(Y)} \tag{8}$$

The use of the Dempster-Shafer theory for information fusion of sensory data was introduced in by Garvey et al. [15]. The theory is more flexible than Bayesian Inference for it allows each source to contribute information with different levels of detail.

3 Related Works

Recent studies about opportunistic collaborative applications in mobile devices generally point to opportunistic sensing, in order to generate information that can be used for posterior analysis.

Cunha et al. [16] presented a case study in which participants used mobile phones as a solution to enable users to opportunistically capture academic presentations collaboratively. In another study, Castro et al. [7] presented a collaborative approach to opportunistic sensing, facilitating gathering of behavioral data from mobile phone users. Most of these approaches focus on generating content from the sensing capabilities of the smartphones, while DMEK focus on sharing content that users themselves generated.

The mobile social software (MoSoSo) based on a P2P network is an application that allows users to discover, communicate and share resources with one another [17]. In another study, Chen et al. [18] presented a wireless real-time music sharing application that lets users play music directly from their smartphones. All users in the vicinity can observe the collaboratively formed playlist on their smartphones in real-time. Even if these works are similar to DMEK, the opportunistic aspect is absent and consequentially these approaches are dependable on internet connections to properly work.

Fortino et al. [19] presented a development methodology for IoT services in heterogeneous environments. Even if their approach could model a service such as MEK, it remains unclear if their profile matching would be an improvement, as the use case presented in the paper is a crowd safety opportunistic IoT Service. In such scenario, the information is processed from a data center, matched and only then sent to users. Although DMEK is more specific in purpose, it does not depend on existing infrastructure to function and relies only on the wireless capabilities of the devices themselves.

4 DMEK: Improving Profile Matching

Our goal is to improve the decision process of MEK to perform exchanges based on profile score similarity. We could view users, in an analogous way, as different sources of information and combine their profile data to obtain scores of similarity and thus address the uncertainty, meaning which profiles should or should not exchange knowledge.

In this work, we propose an extension of MEK that uses a decision system based on a data fusion technique, called Dempster-Shafer (see Sect. 2.3) to integrate interests from different peers into a consistent, accurate, and useful representation of their shared interests in order to determine a good peer to trade

---

Algorithm 1. Decision and Exchange

```
1: procedure EXCHANGE(Users)
2:     for actual, other in Users do
3:         if actual in proximity of other then
4:             sim, dif =
5:             DempsterShafer(actual, other)
6:             if sim > dif then
7:                 dft =
8:                 (actual.docs − other.docs)
9:                 actual.docs.insert(dft)
10:             end if
11:         end if
12:     end for
13: end procedure
```

---

information. We classify if peers are similar based on their scores and determine if the application will exchange knowledge. Considering two users with interests (ranked between -1 for dislike and 1 for like), we obtain the evidences of similarity e_s and difference e_d by taking the rankings provided by the users and calculate their euclidean distance. We denote the profile as a vector of interests rankings $\boldsymbol{R} = (R_1, R_2, \ldots, R_n)$. Let X_R and Y_R be users denoting their respective ranked interests. We obtain their euclidean distance, as the evidence for similarity e_s and compute the evidence of difference by doing $e_d = 1 - e_s$. According to the rule of combination, the resulting similarity is compared to the difference. If the similarity is higher, the users are acceptable to exchange knowledge.

Algorithm 1 illustrates the decision of the application. Each user has a list of interests, and a list of documents. Once in proximity of one another, their evidences of similarity and difference are computed and combined. If similar, their interests are matched and related content is transferred.

5   Implementation and Experimental Evaluation

In this section we describe the experiments with the improved profile matching approach to evaluate the following properties: (I) the execution times of the proposed approach; (II) and the frequency of exchanges of each method, to establish the resource consumption and therefore the impact on the mobile devices. In order to evaluate our approach we performed the experiments by means of a computational simulation.

5.1   Implementation Details and Scenario

The simulation was developed in Python 2.7 programming language with NetworkX package, which provides data structures to represent graphs and algorithms and calculate network properties and measures (such as shortest paths

Algorithm 2. Arrival's distribution of users in the simulation

```
1: procedure ARRIVALS
2:     arrivals = Poisson(keyLocations)
3:     P_A = Probabilities of arrivals
4:     N_A = Length of arrivals
5:     orgns = random(N_A, arrivals, P_A)
6:     dtns = random(N_A, remainingLocations)
7:
8:     for a_r in RANGE(arrivals) do
9:         start = orgns[a_r]
10:        target = dtns[a_r]
11:        path(mapGraph, start, target)
12:        Users.insert(user(path))
13:    end for
14:    return Users
15: end procedure
```

like Dijkstra and A* search). The simulation itself is mainly composed of three procedures, being those: arrivals, movement and exchange.

As seen in Algorithm 2, P_A are the probabilities of arriving at each one of those positions and N_A is the length of the arrivals list, used to maintain the number of origin points correspondent to the number of destination points.

Algorithm 3. Simulation Procedures

```
1: procedure SIMULATION(Interests, Docs)
2:     for step in Duration do
3:         Users = ARRIVALS(λ)
4:         for user in Users do
5:             user.MOVE(θ)
6:             EXCHANGE(Users)
7:         end for
8:         for user in Users do
9:             if user.path = finished then
10:                REMOVE(user)
11:            end if
12:        end for
13:    end for
14: end procedure
```

The paths are lists of coordinates (pairs of latitudes and longitudes) between a starting point and a target, where users travel in the simulation. However, only hopping them through their path's coordinates does not achieve realistic human movement. Therefore user move as the following: (I) First, for an user in the simulation we pick it's current position (i), as well as the next position (n_i), in the path; (II) Then we calculate the bearing between points i and n_i.

(III) With the bearing, we calculate a new point n_p between i and n_i, with a distance of 1.3 m. (IV) Finally, we obtain the distance (in kilometers) between n_p and n_i by using the haversine function. If the distance between n_p and n_i is negligible, we set n_i as the current position of the user and restart the routine from step I. If not, we repeat steps II to IV until the user reaches the next position in the path. By using this approach commuters travel their respective paths in segments, by approximating their current to their next position at each step of the simulation. Figure 3 illustrates this process. After the movement is executed, the exchanges of knowledge are conducted (refer to Sect. 4). Time in the simulation is divided in discrete units called steps, and as parameters it receives all the interests and documents that will be available to users. As they commute along on their paths, when in proximity they exchange information. After the three procedures are executed, users that completed their paths are removed. Algorithm 3 summarizes all simulation stages.

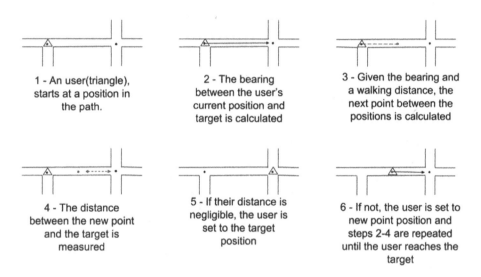

1 - An user(triangle), starts at a position in the path.

2 - The bearing between the user's current position and target is calculated

3 - Given the bearing and a walking distance, the next point between the positions is calculated

4 - The distance between the new point and the target is measured

5 - If their distance is negligible, the user is set to the target position

6 - If not, the user is set to new point position and steps 2-4 are repeated until the user reaches the target

Fig. 3. Demonstration of an user moving in a single path segment. Walking distance is set to 1.3 m and a distance considered negligible is approximately of 10 m

5.2 Scenario

To obtain realistic results, the simulation happens in the actual topology of the city center of Rio de Janeiro, as seen in Fig. 4. The area lies on the plains of the western shore of Guanabara bay and represents the financial heart of the city, comprehending an area of approximately 6.47 km^2, a core point of the central region. Despite having a large number of residences, the neighborhood is predominantly commercial with a mixture of historical buildings, a scenario where many people commute in their daily routines.

5.3   Metrics

We evaluated three methods for profile matching: MEK profile matching, were
no similarity scores are calculated; profile similarity through euclidean distance;
and the improved approach with Dempster-Shafer. We refer to them as NSS
(No Score Similarity), EDS (Euclidean Distance Similarity) and DSS (Dempster-
Shafer Similarity) respectively.

The following metrics were used in the experiments for evaluating the results
of our improved profile matching: The execution times for each method, the
number of exchanges for each method and the corresponding precision and recall
at each step of the simulation.

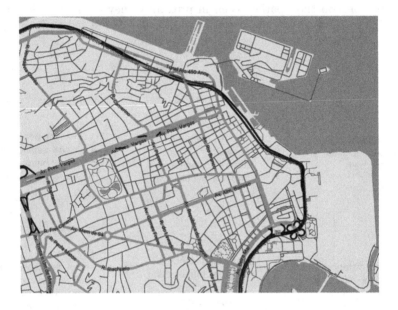

Fig. 4. Topology of the center of Rio de Janeiro

5.4   Evaluation Results for Profile Matching Methods

In our simulation, each user randomly receives interests with corresponding documents. Each document has a size of 3 MB, reflecting the average size of an
music or image file.

Figure 5 shows the amount of exchanges for each method. As we can see, the
amount of data exchanges for NSS method increases highly whereas the other two
methods that calculate profile similarity scores have less data exchange between
the peers, reflecting in lower curves in the graph. These results also highlight that
the effort imposed by NSS is much higher. Since peer discovery and information
retrieval in adhoc networks can be very energy consuming [20]. In this sense the
EDS and DSS methods proved to be more effective.

As seen in Fig. 6, DSS method did not perform better than EDS in precision performance, as the two methods were tied at approximately 98% precision

followed by NSS at 90%. However, this indicates that the exchange strategy
of MEK was not affected by the decision mechanisms while achieving a lower
number of exchanges. The same cannot be said about the recall performance.
As seen in Fig. 7, the NSS method kept a recall of 62% approximately, while
DSS kept at 53% followed by EDS with 45%. This result was expected, since
the decisions mechanisms, although able to keep reasonable precision, lowered
the exchanges numbers impacting directly on the capacity of the application of
retrieving information.

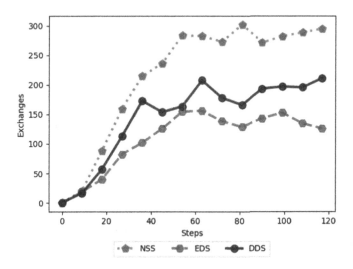

Fig. 5. The average number of exchanges per simulation step

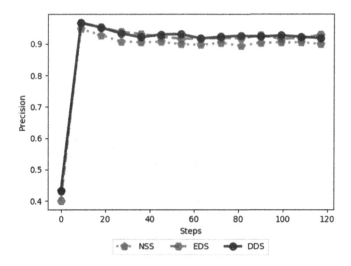

Fig. 6. The precision of each method per simulation step.

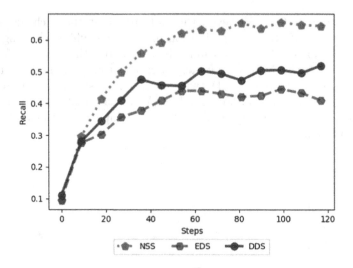

Fig. 7. The recall of each method per simulation step.

5.5   Execution Times

Determining profile matching in MEK must be done in a timely manner. It is important to evaluate the performance of each method regarding their time cost. Therefore, in this section we present the time results, as seen in Fig. 8. DSS times were approximately 2.6 s while EDS times kept bellow 2.4 s, which was expected

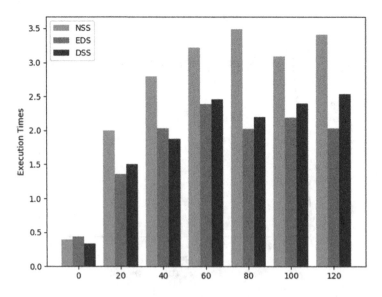

Fig. 8. Time costs for each of the methods used in the simulation.

since Dempster-Shafer performs further calculations than Euclidean Distance. Thus, these results are satisfactory for a real-time solution such as MEK. The greatest advantage of DSS over EDS is that it does not deal with uncertainty, while DSS does. In this sense the close results between DSS and EDS shows DMEK is not only feasible, but presents an actual advantage over MEK and the other traditional profile matching schemas.

6 Conclusion

This paper presented DMEK (Decision Mobile Exchange of Knowledge), a solution where mobile devices are used to disseminate knowledge among it's users opportunistically by using a decision mechanism based on Dempster-Shafer theory of evidence. Our experiments showed that DMEK has a low decision delay while keeping reasonable precision. Even though the recall of the solution has decreased, DMEK has no greater resource consumption compared to the traditional approach. The experiments show the run-time of Dempster-Shafer to be close to other measures of similarity and in this sense we can conclude that DMEK is an improvement.

In future works, we intend to evaluate DMEK in a real-world scenario in order to accurately understand its strengths and weaknesses. Moreover, we will integrate other informations from the devices, such as walking speed of the user and battery level, to support our solution as a way of improving the profile matching. Thus, improving our decision mechanism to perform a better evaluation of similarity. We also intend to experiment other decision techniques/algorithms in the core of DMEK. We envision DMEK as the core of a social routing algorithm to be used in mobile or content-driven networks. This way content would be routed to people based on their social circles and as well as their interactions and characteristics. Finally we intend to add application requirements and QoS constraints into consideration in the future.

References

1. Statista: Forecast of mobile phones worldwide. https://www.statista.com/statistics/274774/forecastof-mobile-phone-users-worldwide Accessed 18 Mar 2016
2. Farias, C., et al.: Multisensor data fusion in shared sensor and actuator networks. In: 2014 17th International Conference on Information Fusion (FUSION), pp. 1–8. IEEE (2014)
3. Atzori, L., Iera, A., Morabito, G.: The Internet of Things: a survey. Comput. Netw. **54**(15), 2787–2805 (2010)
4. Guo, B., Zhang, D., Wang, Z., Yu, Z., Zhou, X.: Opportunistic IoT: exploring the harmonious interaction between human and the Internet of Things. J. Netw. Comput. Appl. **36**(6), 1531–1539 (2013)
5. Kortuem, G., Schneider, J., Preuitt, D., Thompson, T.G.C., Fickas, S., Segall, Z.: When peer-to-peer comes face-to-face: collaborative peer-to-peer computing in mobile ad-hoc networks. In: Proceedings of the First International Conference on Peer-to-Peer Computing, pp. 75–91. IEEE (2001)

6. Savaglio, C., Fortino, G.: Autonomic and cognitive architectures for the Internet of Things. In: Di Fatta, G., Fortino, G., Li, W., Pathan, M., Stahl, F., Guerrieri, A. (eds.) IDCS 2015. LNCS, vol. 9258, pp. 39–47. Springer, Cham (2015). https://doi.org/10.1007/978-3-319-23237-9_5

7. Castro, L.A., et al.: Collaborative opportunistic sensing with mobile phones. In: Proceedings of the 2014 ACM International Joint Conference on Pervasive and Ubiquitous Computing: Adjunct Publication, UbiComp 2014 Adjunct, pp. 1265–1272. ACM, New York (2014)

8. Souza, D.D.S., Silveira, P.C., Oliveira, J., Fogaça, G., de Souza, J.M.: Estudo da aplicação de uma abordagem de disseminação oportunística de dados no cenário de cidades inteligentes (2012)

9. Oliveira, J., Souza, D.D.S., de Lima, P.Z., da Silveira, P.C., de Souza, J.M.: Enhancing knowledge flow in a health care context: a mobile computing approach. JMIR mHealth and uHealth 2(4), e17 (2014)

10. Bluetooth: Bluetooth Specifications. https://www.bluetooth.com/specifications Accessed 18 July 2018

11. Han, Q., Cho, D.: Characterizing the technological evolution of smartphones: insights from performance benchmarks. In: Proceedings of the 18th Annual International Conference on Electronic Commerce: e-Commerce in Smart connected World. ACM (2016). Article No. 32

12. Dempster, A.P.: Upper and lower probabilities induced by a multivalued mapping. Ann. Math. Statist. 38(2), 325–339 (1967)

13. Shafer, G., et al.: A Mathematical Theory of Evidence, vol. 1. Princeton University Press, Princeton (1976)

14. Provan, G.M.: The validity of dempster-shafer belief functions. Int. J. Approx. Reason. 6(3), 389–399 (1992)

15. Garvey, T.D., Lowrance, J.D., Fischler, M.A.: An inference technique for integrating knowledge from disparate sources

16. Cunha, B.C., Uscamayta, A.O.M., Pimentel, M.D.G.C.: Opportunistic recording of live experiences using multiple mobile devices. In: Proceedings of the 22nd Brazilian Symposium on Multimedia and the Web, Webmedia 2016, pp. 99–102. ACM, New York (2016)

17. Tsai, F.S., Han, W., Xu, J., Chua, H.C.: Design and development of a mobile peerto- peer social networking application. Expert Syst. Appl. 36(8), 11077–11087 (2009)

18. Chen, Z., Yavuz, E.A., Karlsson, G.: Demo of a collaborative music sharing system. In: Proceedings of the Third ACM International Workshop on Mobile Opportunistic Networks, MobiOpp 2012, pp. 77–78. ACM, New York (2012)

19. Fortino, G., Russo, W., Savaglio, C., Viroli, M., Zhou, M.: Modeling opportunistic IoT services in open IoT ecosystems. In: 17th Workshop From Objects to Agents WOA, pp. 90–95 (2017)

20. Baumgart, A.S., Knapp, H., Schader, M.: Mobile semantic data exchange in ad hoc networks using distributed profiles. In: 2006 1st International Symposium on Wireless Pervasive Computing, p. 6-pp. IEEE (2006)

Author Index

Printed in the United States
By Bookmasters